Language, Science, and Structure

[L]anguage is a biological system, and biological systems typically are "messy", intricate, the result of evolutionary "tinkering", and shaped by accidental circumstances and by physical conditions that hold of complex systems with varied functions and elements.

—Noam Chomsky, 1995 *The Minimalist Program*

The metaphysical import of successful scientific theories consists in their giving correct descriptions of the structure of the world.

—Ladyman & Ross, 2007 *Everything Must Go*

Language, Science, and Structure

A Journey into the Philosophy of Linguistics

RYAN M. NEFDT

OXFORD
UNIVERSITY PRESS

OXFORD
UNIVERSITY PRESS

Oxford University Press is a department of the University of Oxford. It furthers
the University's objective of excellence in research, scholarship, and education
by publishing worldwide. Oxford is a registered trade mark of Oxford University
Press in the UK and in certain other countries.

Published in the United States of America by Oxford University Press
198 Madison Avenue, New York, NY 10016, United States of America.

Library of Congress Control Number: 2023930572

ISBN 978–0–19–765309–8

DOI: 10.1093/oso/9780197653098.003.0001

Printed by Integrated Books International, United States of America

For Anastasia,
Моя украинская радуга

Contents

Preface

This book is ambitious on a number of fronts. As its primary objective, it endeavours to offer novel insights into one of the most complex questions of human existence, i.e., what is language? The primary method of illumination is the investigation of the best tool in our current arsenal, the science of linguistics. In search of a unified framework for answering deep philosophical questions, I delve into issues in current linguistic theory, formal language theory, cognitive science, and the philosophies of mathematics and science more broadly.

I advocate a thoroughgoing *ontic* structural realist approach to the many interconnected issues at the intersection between the philosophies of language, linguistics, and science. Where I do venture into metaphysical territory, I adopt an unabashedly naturalistic stance (Ladyman & Ross 2007; Maddy 2007; Ladyman, Ross & Kincaid 2013; Machery 2017). Questions of the ontological status of linguistic items like words, the metaphysical import of formal grammars, or the cognitive structure of language will be pursued in accordance with what our best science has to say on the topic. The science in question will sometimes go beyond linguistics to include biology, cognitive neuroscience, and mathematics.

Many of the debates in the philosophy of linguistics have a history in the rise of generative grammar, often described as a scientific revolution. Thus, foundational positions have taken on a partisan flavour with those representing the various camps holding on to predefined or accepted theoretical allegiances. In the mid-twentieth century feuds were establish along philosophical as well as ideological lines with little in the way of conciliation. However, things have simmered down in recent times, opening up the possibility of fresh dialogue and novel exploration into the nature of the field. In addition, there is a plethora of new results, methods, and theoretical alternatives at our disposal. I swear no fealty to any particular theoretical persuasion in the following text. I hope this offers some solace to those who do. But my arguments will unapologetically favour a philosophy of science perspective on all the issues under discussion.

I do not consider myself a mentalist, platonist, nominalist, or a pluralist. In fact, in the first two chapters I advocate doing away with these unhelpful labels. In their stead, I offer an approach which places the role of linguistic structures at the core of the field, one I believe to be faithful to linguistic practice as well as philosophically fruitful. This is the view I develop in careful detail for the rest of the book in the spirit of scientific philosophy. In doing so, I not only discuss syntax and semantics within generative linguistics but I also present evidence from model-theoretic grammars, distributional approaches, embodied cognitive linguistics, and neural network architectures. There is something in it for the whole family, even if certain cousins have been long estranged.

Naturally, the book aims at those interested in the foundations of linguistics and those who work on the nature of natural language more broadly, both in terms of contemporary and historical issues. This is not to say that arguments I make are not generalisable to other fields, in fact many are drawn directly from them. Nevertheless, I believe that there are both linguistically unique puzzles and claims that are *sui generis* to the field. Despite this belief, the larger aim targets the philosophy of science and specifically the philosophy of the special sciences. I hope that this work can convince scholars working in those areas of the rich terrain that is the philosophy of linguistics.

Ryan M. Nefdt
Cape Town, South Africa

Acknowledgements

I have to start my enumeration of thanks with mention of the funding organisations that have made this research project possible. I am greatly indebted to the Vice Chancellor's Future Leaders 2030 programme at the University of Cape Town, the National Research Foundation of South Africa, and the Oppenheimer Memorial Trust for generously supporting me over the past two years.

Two research fellowships especially provided me with the necessary time, space, and collegiate expertise for the completion of the manuscript. The first, as a fellow at the Center for the Philosophy of Science at the University of Pittsburgh, helped me ground the work in the philosophies of science and cognitive science. The second, at the department of Linguistics and Philosophy at MIT, shored up the linguistics elements of the monograph. I am extremely grateful for both opportunities and to Edouard Machery and Agustín Rayo for making them possible respectively.

Next, I would like to thank audiences at various universities at which I presented parts of this book including MIT's Linguistics and Philosophy department, the New York Philosophy of Language workshop, the Kleiner Lecture Series at the University of Georgia, the work-in-progress group at the Centre for Humanities Engaging Science and Society in Durham, the Department of Philosophy Colloquium at the University of Miami, the workshop on Real Patterns and Cognition at the Sante Fe Institute in California, and the Center for the Philosophy of Science at the University of Pittsburgh. To the last of these I am especially grateful, as they hosted a much needed (for me) manuscript reading group on an early draft of this monograph while I was a fellow there. Without the guidance and insightful criticism of Edouard Machery, Kate Stanton, Kevin Dorst, Alnica Visser, Travis McKenna, and Hannah Rubin I don't think the book would be in the shape it is now.

My intellectual debts go much further, of course. Geoff Pullum has been an inspiration to this work and my general early career path in the philosophy of linguistics. Conversations with him over the years, which have ranged from coffee chats in Braunschweig to lunch in Edinburgh to a stroll along the Potomac in Alexandria, have informed the way I think about the subject, its history and philosophy. He has also read over much of my work and offered necessary critique at important moments.

Very similarly, Zoltán Szabó has been a constant source of support and guidance whenever called upon. Zoltán inspired me to write this book a while ago and has since supported its content and structure tremendously. I've learned so much from his expertise in linguistics and philosophy. Though he didn't tell me how hard it would be to write!

My fellow philosopher of linguistics, Gabe Dupre, deserves special thanks for reading through the entire manuscript and offering invaluable comments in terms of both generative linguistics and the philosophy of science. His comments and corrections on my interpretation of aspects of generative theory were especially useful.

There are so many more people to thank. To the philosophers of science, I thank Steven French who intervened in an early paper on structural realism in linguistics of mine, which led to the avoidance of many errors and hopefully a product he would be happy to accept within his framework. Don Ross read the entire first draft with such close scrutiny that his comments have directly impacted the final version in myriad positive ways. I thank Larry Sklar who first introduced me to the wonders of the philosophy of science in all of its rigour and majesty (at a time when I perhaps wasn't quite ready to appreciate it). And my friend, Harold Kincaid, has provided so many invaluable insights not only into the philosophical goings-on of the present monograph but academia, publishing, and general life. I am so very pleased to have realised that there was a philosopher of such note hiding in the School of Economics at the University of Cape Town.

I thank my colleague Bernhard Weiss for giving a chance to a young scholar from an underrepresented group in philosophy (and academia in South Africa) when I was still very confused about the nature of the field. We have become great friends since then but, thankfully, he has never shied away from telling me when my philosophy needs work. I am a better scholar for it. I'm also grateful to Josh Dever and Patrick Greenough for being excellent mentors during my early career days as well as to my first formal linguistics teacher and MSc supervisor, Henk Zeevat, who took me through the annals of HPSG, Dependency Grammar, Optimality Theory, and computational approaches to language. I also thank Terry Langendoen, Reinhard Blutner, Jarda Peregrin, Jeff Pelletier, Rob Stainton, Louise McNally, Dorit Bar-On, Giosuè Baggio, Yuri Balashov, and Michael Kac for all challenging me to develop my thoughts in numerous ways over the course of writing this book. I owe a debt of gratitude to my research assistant, Matt White, for his fastidious attention to detail, and

Peter Ohlin and Paloma Escovedo at Oxford University Press for all of the help and support on the publishing side of things.

Lastly, I would like thank my family for believing that there was a book in me all along. My mother saw it earlier than anyone, including me. Her unending support drove me forward. And my wife, Anastasia, who is my constant intellectual interlocutor and challenger, I might have written this book but I have yet to win a debate with her! Her brilliance, advice, and support gave me hope that this project could achieve fruition, even when I doubted it myself.

1

Introduction

The philosophy of linguistics is a fascinating field of inquiry which develops insights from an astonishing range of related fields, some obvious and others much less so. Historically, issues of ontology have taken on almost metonymous status. More recently, however, this purview has been extended to include methodological concerns, theory comparison, and interdisciplinary connections. Thus, there are at least two ways in which to access the general topic, what I call the *language*-first model and the *science*-first model. In the former, the fields of philosophical semantics, formal logic, and perhaps first-order linguistics take precedence. In a sense, such an approach amounts to an idea similar to that of 'linguistic philosophy' practised at different times in different ways across the twentieth century and exemplified by the ordinary language philosophy movement, i.e., studying language is one way (perhaps the only way, if you were an adherent of the 'Linguistic Turn') of studying the nature of the things which language describes. I take Jerrold Katz to have been the chief advocate of such a view in linguistics. The central idea is that linguistics offers genuine insights and possibly solutions to philosophical problems (and for him, even settles the longstanding debate between Rationalists and Empiricists in favour of the former) or as he put it, "linguistic theory incorporates answers to significant philosophical problems" (Katz 1965: 590).[1]

The science-first approach, on the other hand, treats linguistics as a science and the philosophy of linguistics as a subfield of the philosophy of science. This is the general approach taken in this book, following in the tradition of Ladyman and Ross (2007), Maddy (2007), and scientific philosophy in general (Ross, Ladyman and Kincaid 2013). Thus, the tools of scientific modelling, debates on scientific progress, realism, and structural realism *inter alia* will inform the general methodology going forward. However, in some chapters the lines will inevitably be blurred. For instance, in Chapter 5, following on from the Chomskyan theme of using the study of language as a conduit to the study of mind (so-called 'mentalism'), I will advocate that the study of linguistic structure can illuminate the ontology of structures themselves or rather a particular kind of biological structure. Indeed, the two approaches

Language, Science, and Structure. Ryan M. Nefdt, Oxford University Press.
© Oxford University Press 2023.
DOI: 10.1093/oso/9780197653098.003.0001

are not incompatible. Looking to linguistics as a means of illuminating philosophical problems is still possible on my approach in the same ways that the philosophy of biology or chemistry might attempt to shed light on problems in metaphysics.

In what follows, I provide a guided tour of the general philosophy of linguistics landscape while presenting the main structural realist project of the current book and the sub-problems it helps to solve along the way. The story that will unfold in the subsequent pages of the various chapters will not only track a historical narrative of the changing concepts of language across the twentieth century and into the twenty-first but also develop a unified structuralist picture of linguistics itself, in line with prominent such accounts in the philosophies of mathematics and science but diverging from them in unique ways in response to the unique subject matter at hand. Necessary technical aspects of linguistics, formal logic, cognitive science, and artificial intelligence discussed in so doing will be explained and never assumed. For now, a brief survey of the terrain.

1.1 The Philosophy of Linguistics

The philosophy of linguistics is a broad field with many distinct avenues of entry. At first blush, one might confuse its scope with that of the philosophy of language. Indeed, the language-first approach realises some version of this possibility. As Soames puts it, the "[p]hilosophy of language is, above all else, the midwife of the scientific study of language, and language use" (2010: 1). He goes on to add that "[t]he central fact about language is its representational character" (Soames 2010: 1). These statements highlight two aspects of the philosophy of language that are relevant to our study. One aspect brought out by the idea that the philosophy of language is a midwife to the scientific study thereof, presumably linguistics, is that the former discipline precisifies definitions, makes distinctions, and generally clears the conceptual ground for linguistic theory. Indeed this sounds similar to the remit of the philosophy of linguistics or what it *should* be doing at least. The second claim, that the central feature of language is its representational nature, underlies much of the focus on meaning, be it semantic or pragmatic, in the philosophy of language. Here there is a clear departure from the philosophy of linguistics, as I present it. For instance, in generative linguistics, syntax or the computational system takes precedence over other systems and thus the philosophy of generative linguistics typically involves reflection on that aspect of the theoretical linguistics (as is witnessed in Ludlow's (2011) titular contribution to the subject). Moreover, syntax, phonology, morphology, semantics, and the interfaces between them are all the proper subject matter of the philosophy of linguistics within and beyond the generative tradition. In addition, issues of the psychological reality of grammars, the significance of psycholinguistic evidence in linguistic theorising, and the mathematical underpinnings of different formalisms all form a natural part of our study but seem to be outside of the scope of the philosophy of language proper.

Put in another way, the philosophy of linguistics is the philosophical investigation of a particular kind of scientific theory, namely a linguistic theory. As philosophers of linguistics we are interested in language only *indirectly* as it pertains to how we study and model it. A helpful analogy might be the relationship between the philosophy of physics and metaphysics. The former studies physical reality only through an investigation of the science that describes it. Conversely, metaphysics is no more the 'midwife' of physics than the philosophy of language is that of linguistics. Of course, this is a criticism of analytic or 'neoscholastic metaphysics' within naturalised metaphysics (Ladyman and Ross 2007) and one I endorse fully in this book. In fact, if there is any midwifery going on, it will be done *via* the present methodology.[2] In essence, the position I take amounts to a commitment to *naturalised philosophy of language*.

Of course, no philosophical boundary is completely without exceptions. The current state of metasemantics or the philosophy of semantics lives in both the worlds of the philosophy of language and that of linguistics. The nature of reference, the mechanisms which ground our meaningful use of language, and the implications of mathematical formalisms such as those involved in model-theoretic semantics are central to this emerging field, as they should be. Chapter 8 of this book delves into those issues and attempts a unification with the claims made about syntax and phonology in other parts.

This is all to say that the underlying philosophical framework of the present research project should be kept distinct from that of more traditional approaches to language in philosophy, especially where those approaches did not confirm to the naturalistic stance taken here and elsewhere. In the next section, I will further exemplify the particularities of the kinds of questions that the philosophy of linguistics asks, before giving an overview of the book to come.

1.2 Generative and Non-Generative Frameworks

Generative linguistics is still by and large the dominant force in theoretical linguistics. Naturally, no simple characterisation of such all-encompassing scientific paradigm can be readily provided at the outset. In fact, much of the present work is devoted to providing precisely such an account of linguistic theory in general. Nevertheless, much like relativistic physics or even something more amorphous like political realism in political theory, there are core tenets around which most practitioners operate. In this section, I'll briefly outline the basic tenets of generative grammar and use these as mechanisms for distinguishing between offshoots of the research programme and non-generative frameworks. However, Chapter 2 will revisit the ontology of generative linguistics, while Chapters 4, 5, and 7 will further develop the view in terms of its scientific underpinnings, treatment of individual lexical items (as well as those of other frameworks), and historical roots in the overarching cognitive scientific project.

There are five core characteristics of the generative programme in linguistics which I want to highlight here.[3] I should note that these characteristics do not form a necessary and sufficient set or definition at the level of research tradition but rather a guide to the theoretical underpinnings of subsequent research programmes and how different specific theories developed from core theories or propositions in the sense of Kuipers (2007).[4] The consequence of an exhaustive characterisation of generative grammar, in terms of (1)–(5) below, would be tantamount to the exclusion of the path-breaking *Syntactic Structures* from the generative tradition (since the formalist leanings of that work didn't adhere to (2), (3), or (5) explicitly). But we'll see as the book continues that there is a significant theoretical shift which occurred between 1957's formal instrumental model of syntax to the mentalism of *Aspects of the Theory of Syntax* (1965) and beyond for the study of language in general.

The five properties are as follows:

1. Autonomy of Syntax (or interpretivism): The methodological posit that the core 'generative' component in natural language production is the computational system which produces the set of grammatical expressions. This system operates independently of the semantic, pragmatic, and phonological components of the grammar.
2. Universal Grammar: The claim that despite surface differences between the world's languages, there is a set of genetically endowed linguistic universals common to all possible human languages (though developments

such as the *Principles and Parameters* framework allow for external linguistic input to shape the initial settings of the grammar).[5]

3. Innateness Hypothesis: A rationalistic approach to natural language acquisition in which human infants are endowed with a linguistic system prior to encountering any input (often motivated by the 'poverty of stimulus' argument which is itself a special case of the underdetermination of theory by data argument, in this case the theory is language acquisition).

4. Competence-performance distinction: Linguistic theory is concerned with an ideal linguistic competence and not necessarily with the various aspects of performance or actual parsing and processing in real-time.[6]

5. Rule-based Representationalism: This is the view that the posits of the grammatical theory or rules of the grammar are actual features of the human agent or 'cognizer' (actual goings-on in her mind/brain) at some level of deep neurophysiological embedding. To 'know' or have a language on this view is to have subconscious (tacit or implicit) access to these rules.[7]

For a flavour of the kinds of objections that have been levelled against generative grammar over the years. Jackendoff (2002, 2007) and his parallel architecture theory of language diverges from (1) or the autonomy of syntax, Devitt (2006) mounts a philosophical critique of (5) or representationalism, Cowie (1999) classically questions (2) and (3) (while Putnam (1967) focused mostly on (3)), and more recently empirical linguists and psycholinguists such as Christiansen and Chater (2016) challenge (4) directly.

To see how other frameworks compare to generative grammar requires an appreciation of the more formal mathematical side of the enterprise. This is in part due to the legacy of *Syntactic Structures* and earlier work in formal language theory. Developments on the generative grammar side were heavily influenced by the proof theory, recursion theory, and mathematical logic of the time. For now, we can can understand a generative grammar as a rule-bound device which takes a finite set of rules over a finite vocabulary to deliver a discretely infinite output (much more detail will be added in Chapter 4). There are various ways in which this idea has been instantiated over time in the generative tradition from simple rewriting systems (of the form replace X with Y) with transformations to more set-theoretic functions like *move* and *merge* (the topic of Chapter 7). The central idea was and has always been that a grammar *recursively enumerates* the set of well-formed formulas or in the case of natural language, the set of grammatical expressions.

In a sense, grammars viewed as 'generative' derive (or 'generate') expressions. But this isn't the only way of characterising natural language syntax (and/or semantics). Geoffrey Pullum and others have advocated model theory as an alternative basis for syntax for many years now (it is already a dominant foundation in formal semantics as we will see in Chapter 8). Here the idea is that "[a]n MTS [model-theoretic syntax] grammar does not recursively define a set of expressions; it merely states necessary conditions on the syntactic structures of individual expressions" (Pullum and Scholz 2001: 19). So instead of deriving well-formed formula from rules (as in generative syntax), you stipulate conditions for grammaticality. There are some interesting philosophical consequences of such a move. For instance, it allows for grammaticality to be gradable as opposed to binary (i.e. sentences aren't just in or out of the set but more or less grammatical in terms of the conditions). There are many grammar formalisms that are amenable to such characterisation, they are sometimes called constraint-based grammars (e.g., Generalized Phrase Structure Grammar, Head-driven Phrase Structure Grammar, Arc-Pair Grammar, etc.), where the constraints act like axioms or conditions on grammaticality.

Although generative and model-theoretic (also called 'declarative' or constraint-based) frameworks are often pitted against one another, they are not incompatible. In fact, there are hybrid cases involving both generation and constraints like Lexical Functional Grammar (LFG) and Optimality Theory (and perhaps even Government and Binding theory). For example, LFG employs two distinct levels of syntactic structure, the *c-structure* which is a phrase-structure or generative representation and the *f-structure* which treats grammatical functions in similar ways to the ways in which constraint-based approaches do. The latter representation resembles yet another non-generative approach to grammar in Dependency Grammar which doesn't fit into either the category of model-theoretic or generative neatly (the details of which we will encounter throughout the book). Moreover, there are formal proofs which prove various results about weak and strong equivalence between all of the above. Needless to say, it's enough to keep a philosopher of science very busy! It will certainly keep this philosopher of science busy for much of the book.

To compare model-theoretic accounts with generative-enumerative ones is to embrace the formalist side of linguistics. Focusing on the less formal side opens up comparison classes between generative grammar and cognitive linguistics. The latter framework adheres to the idea that linguistics is a member of the cognitive sciences club or what George Lakoff (1991) calls the 'cognitive commitment' but it rejects both the claim that generative grammar truly incorporates this ideal and, with later instantiations, that linguistics is the core subject in the study of mind. I challenge the intensity of these claims towards the end of the book while exploring and utilising various aspects of cognitive linguistics in the philosophical picture throughout (see Chapter 6 for insights from construction grammar in the ontology of words).

For now, however, I would like to move on to some of the ontological and scientific tools the book will use to make sense of the language sciences construed broadly enough to meaningfully include most of the frameworks mentioned above, a special goal of this particular volume.

1.3 Structures and *Structuralisms*

To speak of 'structuralism' in philosophy is to make reference not to a single theory but a family of theories with distinct aims, targets, and features. Even the term 'structural' is polysemous if not vague at times. Thus, attempting to chart this course can feel like a downhill slalom. But chart it we must! Especially if both ontology and the philosophy of science are within our remit. I will briefly discuss the senses I make use of in this text, mostly obviating the multitude of other meanings outside of philosophy, for instance, in psychology (although touched upon in Chapter 9), taxonomic linguistics (mentioned again in Chapter 2), and economics (quite a distance beyond the current purview). Structuralism has a long and developed history in linguistics, Ferdinard de Saussure's (1916) *Cours de linguistique générale* set the stage for modern linguistic thinking. In that work, the now famous distinction between *langue* which is a system of signs in which each (mental) sign is the conjunction of two values and *parole* or the utterances and texts that individuals produce and/or understand when they use *la langue*. One central theoretical claim from this framework concerns the 'arbitrariness of the sign'. For Saussure, this is the 'first principle of the linguistic sign' that despite possible mimetic links between signifier (e.g., words) and signified (or objects they denote), "the choice which calls up a given acoustic slice for a given idea is perfectly arbitrary. If this were not so, the notion of value would lose something of its character, since it would contain an element imposed from without. But in fact the values remain entirely relative, and that is why the link between the idea and the sound is radically arbitrary" (1916: 157). Importantly, for structuralists of this variety, the 'signified' is not the referent as might be assumed in the philosophy of language. Rather "each signifier and each signified is a value produced by the difference between it and all the other signifiers and signifieds in the system" (Joseph 2017: 7). Thus, Saussure's linguistic structuralism is the view characterised by the claim that language is a system of values induced by elementary oppositions (such as the voiced and unvoiced distinction in phonology, etc.) in which particular signs are largely arbitrary. In other words, the relations between symbols and not their particular individual significance is what counts. This much is common to other forms of structuralist thinking.[8]

In the philosophy of mathematics, structuralism has established itself as a dominant paradigm (Hellman 1989; Shapiro 1997; Resnik 1997; Chihara 2004). According to one of its most prominent advocates, Stewart Shapiro, the trend toward structuralist thinking in meta-mathematics stems from Hilbert's axiomatics and the introduction of implicit definition (see Shapiro 1997, 2005). Mathematical structuralism is thus the view that mathematics is the science of abstract structures, where the objects in these structures are defined purely with relation to each other and the overarching superstructure. On this view, mathematical objects have no internal or individual natures in themselves. Different structuralist proposals differ on what they take these structures to be exactly, eliminative or object-preserving, and what the background theory looks like, modal logic or set-theoretic, first-order or higher-order etc. The metaphysical import of this approach is that mathematical theories are *about* structures and not individual objects. This view was in part developed in the wake of a problem famously identified for Platonism by Paul Benacerraf (1973). The dilemma posed by Benacerraf makes the claim that the quest for mathematical truth pulls in two opposing directions with relation to a uniform semantics and a (causal) epistemology.[9] In other words, the more our theories refer to abstract (non-spatio-temporal) objects in the semantics, the further away a tractable epistemology is from our grasp. As we will see, the issue with Platonism in linguistics is similar in that its abstract ontology is at odds with our abilities to *know* or more pertinently to learn languages.

In Chapter 4, I aim to unite the formal structuralist perspective in the philosophy of mathematics with a real pattern analysis (Dennett 1991) of natural language structure.

In the philosophy of science, structuralist thinking was developed in response to a different set of problems. The basic idea, to be revisited in more detail in Chapter 7, is that we as realists about scientific truth are caught between the pull of realism and the rational scepticism of anti-realism. Structural realism has often been considered a happy medium (the 'best of both worlds' strategy in Worrall's words).

It is well-known that traditional scientific realism in the philosophy of science faces a serious challenge often referred to as pessimistic meta-induction or the problem of radical theory change. This problem relates to explaining progress in science and the ever-changing reference of scientific terms. If our theories are true of the world (or even approximately so), then how can we explain scientific progress in cases in which theories have been radically altered (as in the move from Newtonian to relativistic physics or quantum theory)?

One answer to these sorts of worries is scientific anti-realism. On views under this framework, scientific theories need only be empirically adequate (get the 'observables' right). Van Fraasen (1980) is one prominent example of this kind of view. Interestingly, this latter work has led to much of the focus on modelling in contemporary philosophy of science (which we will consider in both Chapters 7 and 8). Although this might be a viable option, it does lead to similar worries to that of instrumentalism in rendering the success of our models or grammars inexplicable (or miraculous).[10]

There is, however, a more modest alternative in views under the banner of structural realism. As Ladyman puts it,

> Rather we should adopt the structural realist emphasis on the mathematical or structural content of our theories. Since there is (says Worrall) retention of structure across theory change, structural realism both (a) avoids the force of the pessimistic meta-induction (by not committing us to belief in the theory's description of the furniture of the world), and (b) does not make the success of science [...] seem miraculous (by committing us to the claim that the theory's structure, over and above its empirical content, describes the world). (1998: 410)

These issues, I argue, have direct relevance to the philosophy of linguistics, where radical theory change and an abundance of alternative models can also be found. Viewing science and, in particular, language science along the *ontic* structural realist lines of Ladyman (1998), Ladyman and Ross (2007), French (2014) among others will thus be a central theme of the book. Specifically, I follow the latter authors in rejecting the epistemic variant as inadequate for the task of dismissing the metainduction and explaining novel predictive successes.

There is also a more metaphysical side of structuralism exemplified by the recent work of Ted Sider (2013, 2020). For him, "the idea is that patterns or structure are primary, and the entities or nodes in the pattern are secondary" (Sider 2020: 3). He distinguishes between three versions of structuralism: *nomic essentialism*, *comparativism about quantities*, and *structuralism about individuals*. The last of these options is connected to the general views found in structural realism and mathematical structuralism. I won't be discussing the more metaphysical concepts of bundle theory or Sider's (2013) notion of 'fundamentality' as the true subject of metaphysics since my focus is not on pure metaphysical considerations here. This is not a work in metametaphysics, but rather in the philosophy of a particular special science. However, in my own way, I endorse and develop his 'primitivism' about structure and the importance of the theoretical inferential role structure plays in explanation in linguistics.

More generally, it is the perspective that structuralism brings with it that will be harnessed as a philosophical strategy in the subsequent pages. I take this strategy to involve both abstracting away from individual 'object-oriented' ontologies in asking metaphysical questions and homing in on transmittable structural features in asking scientific ones. Thus, my thinking and arguments are certainly structuralist in persuasion but not identical to either the strategies employed in the mathematical structuralist or structural realist literature.

1.4 A Guide to the Book

The overall plan of the book is to take on the task of providing an account of language as a structure or pattern and linguistics as a science of that structure. The connecting line will be a development of Dennett's information-theoretic idea of a *real pattern*, modified by Ladyman and Ross (2007) and others on the ontological side. The theoretical side will be captured by ontic structural realism and its promise of unification across the different subdisciplines of linguistics as well as the cognitive sciences. Each chapter either sets the stage for this view or provides additional evidence and argument for the appeal of these twin goals over certain rivals.

In terms of stage setting, Chapter 2 comes closest to addressing the traditional foundations debate on ontology and the subject matter of linguistic theory. But I carve the partitions in distinctive ways from those older discussions. For instance, I divide different foundational views in terms of a novel taxonomy involving those committed to objects and those committed only to states or networks. However, I distance myself from the usual way of representing this debate in terms of the standard metaphysical categories of *platonism*, *nominalism*, and *conceptualism* (Katz and Postal 1991). Rather I adopt a perspective based on the kinds of things theorists take language to be, thereby tying the metaphysics to the *realist* study of language. Thus, platonist views that take languages to be individual (platonic) objects are conceptually connected to nominalist views which involve the same presupposition—what I call 'object-oriented' views. These are contrasted with state or network views, which identify languages not as objects but rather as states of interconnected properties (the classical generative example being the *Principles and Parameters* framework). Besides Chomsky's I-language and Millikan's (1984, 2005) biologically informed conventionalist account, this categorisation also allows me to superficially connect the dominant generative ontology to other approaches such as Labovian socio-linguistics and Brandomian inferentialism.

Chapter 3 deals with two opposing concessionary positions, namely anti-realist views (the later-Davidson and the right-now-Rey representing this camp) and pluralist views (recently expressed in Stainton (2014) and Santana (2016)). I object to both of these strategies on philosophical grounds. I agree with the sentiment of both global and local anti-realism but consider the *explanandum* to be better accounted for by a specific notion of structure provided in Chapter 4. Similarly, pluralism is meritorious from a number of perspectives, especially methodological ones, but I argue that it invites inconsistencies in terms of ontology. While Chapters 2 and 3 offer unique characterisations and arguments on the extant literature, it is possible for the reader familiar with these debates to skip them without losing the thread of the positive account to come (perhaps glance over the central insights).

Chapter 4 advocates a view that aims to meet certain naturalistic criteria for the study of language without sacrificing realism, coverage or theoretical appeal. This makes it pluralist in spirit but not in ontology. Here, I begin my positive account of what a language is. Specifically, I take languages to be non-redundant real patterns (Dennett 1991) chiefly identified by formal grammars which are essentially compression algorithms. This is related to a view I have championed in many places (Nefdt 2018a, 2019a, 2019b, 2020) without the real pattern analysis. It shares characteristics with the *ante rem* or noneliminative structuralism of Shapiro (1997) which makes room for both realistic talk of objects-in-structures and structures themselves and 'rainforest realism' with sees patterns going 'all the way down' (Ross 2000, Ladyman and Ross 2007).

In Chapter 5, I attempt to ground the previous ontology within a novel space. Specifically, I make the claim (*pace* Chomsky) that 'linguistically real patterns' can be identified not necessarily within individual biological organisms but rather within biological *systems* at the material level. To complete this picture, I suggest the application of this view to language acquisition as pattern recognition. Here, I claim an analogy with neural network architectures in both cognitive neuroscience and machine learning. Neural nets are complex structures which, through connections and weights of those connections, can learn how to pick out real patterns in nature. Linguistic applications abound including machine translation and speech recognition. I go on to show that the way in which such a machine learns a pattern can provide a useful clue into how we as humans can acquire and process linguistic structures.

In Chapter 6, before moving on to a more global linguistic application of some of these ideas couched within a structural realist framework, I focus attention on a sub-problem: what are words? This question has recently attracted a lot of philosophical attention. I aim to show that words are equally susceptible to a pattern-theoretic structuralist characterisation *contra* most of the current accounts on the topic. Chapter 6 will thus be a brief stopover in applying the machinery developed so far before the longer haul flight into new scientific territory.

In Chapter 7, I extend the discussion of structures to the philosophy of science more broadly. In this chapter, the purview changes from direct ontology or scientific metaphysics to the nature of theories and their objects. I develop a structural realist account of linguistics. The idea of this chapter is to present two general reasons for an *ontic* structural realist (Ladyman 1998, Ross 2008, French 2014) interpretation of linguistics. The first is an analogue of the problem of theory change (or 'pessimistic meta-induction', Laudan 1981) through the short history of formal linguistics. The second is to offer a mechanism for theory comparison across frameworks and traditions. Both of these reasons are well-addressed, I argue, within the particular version of structural realism I endorse.

Chapter 8 attempts to add meaning to the formal levels of pattern description so far. The idea is that the key to the progression and connection between different subfields of linguistics is the recognition of systematic patterns within the noise of each subsystem. These phonological, semantic, and pragmatic patterns are also often identified by structural features within a nested network of complex information. The central insight here is an extension of structural realism to some aspects of what is called 'informational structural realism' and the method of levels of abstraction (Floridi and Saunders 2004, Floridi 2008a,b, 2011). This view promises to provide a new look at the linguistic interfaces.

Whereas Chapter 8 ends with a focus on the relationship between syntax and other branches of linguistics, Chapter 9 starts with the questions related to the relationship between linguistics and the cognitive sciences more broadly construed. If some of the earlier chapters can be seen under the lens of the philosophy of science, this final chapter should be viewed in terms of the philosophy of cognitive science.[11] I take a brief detour into the early history of cognitive science in North America in which linguistics was a leader and trailblazer before skipping ahead a few decades to the current *status quo* (with the so-called 'Second Generation Cognitive Science') in which the place of linguistics and the study of language at the centre of the interdisciplinary project is eroding. I caution against this trajectory and show that the overarching philosophical view developed in the present volume can offer a means of integration (or unification) with the cognitive sciences for linguistics and a defence of the salient position of language in the study of mind more broadly construed.

Starting from Chapter 2, each chapter begins with a 'central insight' to be defended. In terms of the dialectic, Chapter 2 comes closest to engaging the traditional topics in the philosophy of linguistics, Chapter 4 is more technical in terms of information theory and formal linguistic theory, while Chapters 7 and 8 focus their attention on the philosophy of science and theory comparison. Chapter 9 attempts to provide an answer the question of the place of linguistics within the broader cognitive sciences anew. Finally, in the conclusion, the eight central insights are laid out as pieces each contributing uniquely to the overall structuralist puzzle of the book.

Thus, in the conclusion, I take a brief step back to assess the overall picture. I also address residual worries, potential objections, and I acknowledge some gaps for further development. I conclude with a review of the book as a whole and where I think its main contributions might lie.

2

Old Landscapes, New Maps

In this chapter, I will apply a fresh perspective on a series of old debates on the ontology of language. In most cases, I find the erstwhile divisions and arguments to be lacking, especially those that rest on *a priori* metaphysical categories such as the type-token distinction. Nevertheless, a number of accounts provided useful tools for further development. Thus, the first central claim or insight of the book, in slogan form, will be a defence of the following negative position:

CENTRAL INSIGHT I: OBJECT-ORIENTED ACCOUNTS OF THE ONTOLOGY OF LANGUAGE ARE NONSTARTERS AND ANY FOUNDATIONAL VIEW WHICH ENDORSES THEM IS THEREFORE LACKING.

We will not yet aim to establish that systems or pattern-based approaches are preferable, except perhaps by elimination. That is the task of the next chapter. Nor does this chapter show that all possible interpretations of platonism, nominalism, and so on are untenable. In fact, I go on to maintain and draw multiple insights from the extant and erstwhile work on linguistic ontology. My main target is object-oriented interpretations of these kinds of philosophical theories—*de dicto* not *de re* instantiations, if you will.

The debate concerning the ontological nature of language achieved prominent status in the philosophy of linguistics during the mid-twentieth century as the science of linguistics was finding its foundations. But philosophical reflection on language dates back to ancient times. Plato, Aristotle, and the Stoics all asked related but distinct questions on this issue. As Itkonen states,

In his dialogue *Cratylus*, Plato (428–348 BC) raises the question concerning the nature of language: does it exist *nomōi* or *phusei*? These are dative forms of *nomos* and *phusis*, respectively. The former means 'law' and 'convention' while the latter means 'nature' in the sense of 'essence'. If language exists *nomōi*, it is conventional or arbitrary. If it exists *phusei*, it is an instrument that has been devised by a mythical or imaginary 'namegiver' (*onomathetēs* or *nomothetēs*)

Language, Science, and Structure. Ryan M. Nefdt, Oxford University Press.
© Oxford University Press 2023.
DOI: 10.1093/oso/9780197653098.003.0002

so as to reveal the essence of its referent; and this it can do insofar as it is a picture of the latter. (2013: 749)

Somewhat prophetically, the distinction Plato makes between convention-alism and essentialism tracks the lines between contemporary inferentialist approaches to language which view language in terms of norm-governed practices and more ontologically committing views like linguistic platonism. Aristotle, on the other hand, emphasised the logical connections to language in terms of syllogisms. His was the first systematic analysis of sentence structure, predicate-argument structure and lexical categories (see the *Categoriae* and *De Interpretatione* respectively for the accounts). Montagovian semanticists too are often associated with the claim that there is no important theoretical difference between formal and natural languages (Montague 1970, Thomason 1974).

The Stoics directed their focus more towards language itself as opposed to its relation to logic (although their propositional calculus was rather devel-oped). Their ontology was thoroughly materialistic. In terms of Plato's earlier distinction, they favoured the *phusis* side. The Stoic philosophy of language involved 'sayables' or *lekta* which resemble contemporary propositions in a number of ways (but differ in aspects such as tense). Although *lekta* were technically incorporeal for them (see Barnes, Bobzien and Mignucci (1999) for a comprehensive account), the rest of their ontology resembled a strongly physicalist theory.

> It's not wrong to say that all existent things are corporeal according to the Stoics, but one needs to add that existent things don't exhaust their ontology. All existent things are, in addition, *particulars*. The Stoics call universals 'figments of the mind' (Baltzly 2018).

In the contemporary setting, the Stoics, like Aristotle, would reject platon-ism about linguistic objects. So languages would not be identified with such entities for them. But they did have a nuanced mentalistic ontology, in which universals were mental in kind and the soul was an mental entity analogous to physical organs. We'll see a similar view within Chomskyan mentalism in Section 2.3.1.

In this chapter, we are going to chart the contemporary terrain on the ontology of language. But we'll need some navigation tools to find our way across the disputed territory. Traditionally, the dominant approach to navigat-ing this landscape has been based on standard metaphysical categories such as *platonism* and *nominalism* with the addition of Chomskyan *conceptualism*. This is the so-called 'Katzian trichotomy' (Katz 1981, Scholz et al. 2022). Unfortunately I think that not only is this tool outdated but it's also a false trichotomy. There are many ontological views in the philosophy of linguistics which are simply not captured by this perspective. In other words, the landmarks Katz' taxonomy capture track metaphysical categories which are not helpful in navigating the contemporary space of possibilities on the ontology of language.

The main reason for the defectiveness of such categories is that they were not inspired by truly naturalistic criteria. I will return to this point in Section 3.3 of the next chapter and develop a more naturalistic account in Chapter 4. Even before such an account is on the table I think it is useful to redraw the lines on the ontology question in terms of more contemporary positions. Thus, the theoretical perspective offered here will focus on which views are committed to objects, networks, or support noncommittal, eliminativist, and of course pluralist ontologies. I will suggest that most of these views have something illuminating to offer to the more structuralist view described in Chapter 4. But by themselves, they fall short of a scientific philosophical position as they often rely on *a priori* analytic metaphysics and domesticated linguistic theory.

For now, we need to be clear about the precise nature of the task ahead.

2.1 What Is a Language, Anyway?

In order to embark on our novel cartographic exercise, we're going to need some new landmarks. Simply asking 'What is a language?' won't do. There are no uncontroversial answers. For instance, if I start by suggesting that languages are primarily communication devices, then I risk alienating those who believe language evolved for the expression of internal thoughts, and that communication is some sort of exaptation. Similarly for the reverse move of identifying languages with their propositional contents and rendering communication ancillary. But within the conflict, then can be conciliation. Of course, natural languages can be both tools for communication and expressions of inner thought contents. Katz' 'principle of effability' comes close to fulfilling both roles simultaneously, effability for him,

[C]omes much closer to expressing the essence of natural language than any proposal so far. In particular [...] even Chomsky's principles of creativity, appropriateness and stimulus freedom, significant as they are to our conception of natural language, fail to put their finger on just what it is that makes a natural language unique [...] only the ability of natural languages to provide a sentence for any thought [...] differentiates them from artificial languages (which gain their expressive power parasitically from natural languages), and from animal communication systems (which, although they can bear amazing resemblance to natural language, are patently non-effable). (1978: 210)

In other words, the hallmark of natural language is that any possible thought can be expressed in any possible natural language. A powerful principle with a simple flaw. If just one sentence S_1 in a language L can be shown not to be translatable into language L', then effability is falsified. Naturally, this latter possibility depends on how we interpret the term 'translatable'. Too local a definition and every idiom would be enough to contravene the principle. Most theorists who have investigated this claim directly (Schnitzer 1982; Malpas 1989) take the global approach and focus on entire languages. Of course, the most obvious opponents of the principle would be linguistic relativists who hold that individual languages represent world views untranslatable into languages which lack the need for those specific conceptual resources. The so-called strong 'Sapir-Whorf Hypothesis' posits that "[w]e see and hear and otherwise experience very largely as we do because the language habits of our community predispose certain choices of interpretation" (Sapir 1929: 209) and that "the world of phenomena we do not find there because they stare every observer in the face; on the contrary, the world is presented in a kaleidoscopic flux of impression which has to be organized by our minds" (Whorf 1956: 212). Different language users carve up nature in distinct and sometimes incompatible ways. Unfortunately, Whorfianism is notoriously short on empirical support and weaker versions such as Boroditsky et al. (2003) are well-supported in terms of cross-linguistic research but do not go so far as to challenge any reasonable account of global intertranslatability.

There are of course other properties which have been offered as constitutive of natural language. Hockett's (1958) famous design features are still standardly touted in linguistics courses across the globe. He focused on aspects such as 'stimulus independence' (or 'stimulus freedom' as Katz calls it) and 'structure dependence'. Chomsky has emphasised the role of unbounded creativity (as alluded to by Katz above), the universality of recursion (we'll return to this issue throughout), and even the design perfection of the linguistic system (Chomsky 1995). Such claims led to a wide-scale scavenger hunt for their analogues or instantiations in other animal species to varying and debatable fecundity. For instance, animal communication, such as bird calls, have been studied in terms of whether they incorporate recursive structures. Gentner et al. (2006) claim that European starlings (*Sturnus vulgaris*) apparently recognise recursive and self-embedding patterns. Corballis (2007) disputes such claims by showing that simple counting strategies are sufficient and that the evidence is inconclusive as to the appreciation of recursion in nonhuman animals. Berwick et al. (2011) apply some nuance and claim that birdsongs only express what they call 'phonological syntax' which is "a formal language; that is, a set of units (here acoustic elements) that are arranged in particular ways but not others according to a definable rule set" (118). One thing most of these discussions have in common is that human language is rule-governed, as this rather lengthy quote from Smith and Wilson illustrates:

> At different times, different features of language have struck people as particularly significant, typical or worthy of attention. Any system as complex as a human language is bound to lend itself to a variety of independent approaches. For example, languages are used to communicate; one obvious line of research would be to compare human languages with other systems of communication, whether human or not: gestures, railway signals, traffic lights, or the languages of ants and bees. Languages are also used by social groups; another line of research would be to compare languages with other social systems, whether communicative or not: economic, political or religious, for example. Again, languages change through time: comparison of languages with other evolutionary systems, organic and inorganic, might also be pursued. While all of these approaches have undoubted appeal, there is an obvious logical point to be made: one must be able to describe a language, at least in part, before going on to compare it with other systems. It seems to us that there is no way of describing or defining a given language without invoking the notion of a linguistic rule. (1985: 325)

Indeed, they are correct that linguistics as a science has held central the concept of a linguistic rule or language as a system of rules. We will follow in this trajectory here. Furthermore, a number of divergent foundational views retain this central feature. Inferentialists, as we will see, dispute the internal rule systems of generative grammarians but maintain that language can be characterised by external normative rules of linguistic communities. Additionally Ruth Millikan (1984, 2005), with her more biologically inspired framework, makes room for rules as conventions passed down from generations to solve coordination problems. Formal linguists and platonists unsurprisingly place formal rules at the centre of their respective characterisations of natural language. In Chapter 4, I will offer an account which similarly places significance on the concept of a formal linguistic rule, but without the baggage of platonism.

What exactly a linguistic rule is will depend partially on the ontological framework under discussion and partly on the working linguist and some notion of 'normal science'. This is the case in a number of sciences in which ontological rumination often operates orthogonally to the work of practitioners whose jobs are to further the scope of theory by means of the available tools on offer. For instance, in our case, a platonist, generative grammarian, and Devittian nominalist could all agree on the grammatical formalism used but dispute its ontological import. In other words, the same grammar could pick out rules in the heads of cognisers, nonspatiotemporal realms, or the output of an idealised community of speakers. For now, we will detail the different ontological positions which have traditionally occupied the landscape on the foundations of linguistics.

In Section 2.2., we'll look at object-oriented views, in Section 2.3, state and network views while in the last two sections anti-realist and pluralist views will be covered respectively before characterising an underlying problem many of these accounts face in terms of a hidden dualism at their core.

2.2 Object-Oriented Accounts

Object-oriented accounts are those which take natural languages to be individual objects of some sort. By this I mean, they take languages to be *ontologically independent* of one another and/or other aspects of cognition. Written slightly differently, English is a separate entity to Kiswahili. Languages are like people; we are often related to one another and can even rely on each other, but when it comes down to brass tacks we are individuals. In Chapter 6, we will make this notion much more precise with relation to the ontology of linguistic items such as words, especially as it is more easily stated within a structuralist framework (Linnebo 2008a). For now the intuitive idea should suffice. The methodological version of this view is often attributed to the American structuralists such as Leonard Bloomfield and Zellig Harris. The advent of generative linguistics is often characterised as a sharp paradigm shift eschewing the tenets of what was known as 'structural linguistics' which came before. Some of these alleged tenets include (1) the limitation to classificatory or taxonomic methods of study, (2) the restriction of data to language corpora (producing only so-called 'observationally adequate grammars'), (3) the rejection of mentalism, and (4) a local limit on language-specific rules and generalisations (i.e., no Universal Grammar).[1]

It is unclear whether or not American structuralism adhered to all of these tenets. In fact, it is unclear whether there was any such unified body of work corresponding to the label of 'American structuralism' in the first place (see Matthews 2001). But one remnant of early anthropological linguistics, and the linguistic relativism with which it was sometimes associated, is the focus on individual disconnected languages. Linguists were explicitly resistant to assuming any significant connection between the grammars of different languages or language families. Studying a indigenous language of Australasian descent can tell us nothing about the structure of English and *vice versa*. In other words, these languages were like ancient artefacts found at different archaeological sites, there was no reason to think that they evinced any sort of underlying or connected picture of human nature or human language (even if they do in reality). An expert in Haryanvi is as distinct from an expert in Nguni as mechanic is from a carpenter, different parts for different uses.

There is an interesting analogy here with American structuralism and van Fraasen's *constructive empiricism*, at least in the latter's extreme empiricist formulation. For him, "[e]mpirical adequacy concerns actual phenomena: what does happen, and not, what would happen under different circumstances" (van Fraassen 1980: 60). Against a realist attitude towards unobservables and objective modality, van Fraasen paints a picture of scientific practice as involving getting the observables right or 'saving the phenomena'. So, "[s]cience aims to give us theories which are empirically adequate; and acceptance of a theory involves as belief only that it is empirically adequate" (1980: 12). The American structuralist concept of 'observational adequacy' tracks a similar practice. However, as Ladyman and Ross (2007) argue, van Fraasen tacitly requires some notion of objective modality to make proper sense of scientific practice. Similarly, understanding language *simpliciter* requires more than just describing grammatical facts *in vacuo* but involves describing grammatical rules. These rules have implications beyond actual corpora, e.g., there are grammatical sentences that have never been uttered. Chomsky realised the need for a deep level of analysis in these terms. Again, we'll discuss that view in the next section.

Interestingly, the move from anthropology to logic and mathematics didn't seem to deter ontological atomism in early linguistic theory. There are at least two distinct but related notions of the term 'formal language' found in the philosophical literature on natural language. Neither forces a departure from the atomistic picture.[2]

In the rest of this section, I will outline two prominent object-oriented accounts which draw respectively from the formal logical and anthropological or social traditions.

2.2.1 P-languages

The formal methods and tools used widely in linguistics, such as recursive grammar formalisms and the linguistic infinity postulate (or the idea that natural language is discretely infinite in magnitude), led to a radical interpretation of the field as an enterprise in line with mathematics or logic. Katz (1981) and Postal (2009) offer Platonism as a grounding for both linguistic ontology and practice.[3] Hence languages are like Platonic objects or P-languages (a term coined by Chomsky (1991)).

Basically the view states that natural languages are abstract objects on the same level as sets or numbers. Linguistics itself is a *sui generis* formal science which describes this particular class of *abstracta*. The usual arguments in favour of this position take the form of criticisms of the received or mentalist position. But we will focus here on the positive aspect of the account, especially with relation to its commitment to abstract objects (with emphasis on 'object').

Specifically, Platonists start with the ontology of sentences and then associate entire languages with sets of those sentences. Katz (1985: 18) states the position in the following ways:

> [G]rammars are theories of the structure of sentences, conceived of as abstract objects in the way that Platonists in the philosophy of mathematics conceive of numbers [...] They are entities whose structure we discover by intuition and reason, not by perception and induction.

And,

> Sentences, on this view, are not taken to be located here or there in physical space like sound waves or deposits of ink, and they are not taken to occur either at one time or another or in one subjectivity or another in the manner of mental events and states. Rather, sentences are taken to be abstract and objective.

Postal (2009) presents a similar argument to this effect. However, he follows Katz (1996) in availing himself of the traditional type-token distinction in metaphysics. He argues that if linguistic theory or grammars were indeed about brain-states or mental events as the biolinguist would have it, then the sentences of these theories would have to be at the level of tokens, not types (which are here conceived of as abstract objects). Soames (1984) makes a similar case for the 'empirical divergence' and 'conceptual distinctness' of linguistics and psychology. Postal states two issues with the received position. For one thing, it seems out of touch with linguistic practice in which grammars usually deal with "island constraints, conditions on parasitic gaps, binding issues, negatively polarity etc." (Postal 2009: 107). Importantly, these accounts are rarely, if ever, informed by evidence from neuroscience or psychology (as one would expect if they were truly concerned with brain-states). Therefore, he concludes that these accounts are concerned with sentence types conceived abstractly.[4]

> Sentence tokens exist in time and space, have causes (e.g. vocal movements), can cause things (e.g. ear strain, etc.). Tokens have physical properties, are composed of ink on paper, sounds in the air [...] Sentences have none of these properties. Where is the French sentence *a signifie quoi?* - is it in France, the French Consulate in New York, President Sarkozy's brain? When did it begin, when will it end? [...] Such questions are nonsensical because they advance the false presupposition that sentences are physical objects. (Postal 2009: 107)

Such considerations lead Platonists to conclude that linguistics is concerned with sentences on the level of abstract objects, in the sense of non-spatio-temporally extended entities. Truth in linguistic theory or in its grammars is then determined by correspondences between the sentences of the theory and these objects. I think this latter leap is not required and, in fact, is rather inimical in light of better options and more naturalistic ontologies, both of which I shall present in this chapter and the next. However, there is some kernel of truth to the notion that linguistic grammars and the theories they inform do possess a formal and abstract level of description through the analysis of sentence types (or whichever type of basic unit with which one begins). Furthermore, a realist account of linguistics should provide an appropriate interpretation of this aforementioned level of abstraction and linguistic practice as it is.

Platonists generally attempt to resituate the debate on the foundations of linguistics in terms of the philosophy of mathematics. However, they failed to appreciate the problems with Platonism in other fields of inquiry, especially mathematics as we will see (Linsky and Zalta 1995; Shapiro 1997; Nefdt 2018a).

Not only this but they do not in fact do justice to the science of linguistics in the ways they claim by insisting on an object-oriented ontology. I will return to these points below but for now let us consider a different object-oriented view which eschews talk of abstract objects altogether.

2.2.2 Devitt's Linguistic Conception

There is a more planetary concept of language which also falls under the auspices of object-oriented views, and it is often associated the notion of an E-language or external language. This time the analogue is not with abstract objects but ordinary physical objects used by real communities of speakers. If Platonism identifies languages with their types, then this view identifies them with their physical tokens. In this sense, languages are subsumed under a folk physics of external objects. There are a few hallmarks of such views and they tend towards ontological nominalism. The chief proponent is Michael Devitt (2006, 2008, 2013).[5]

At the time of Katz's identification of the major ontological positions in the philosophy of linguistics, there was no clearly developed nominalist position. The pre-generative American Structuralist programme of Bloomfield and Harris offered some insights but their anti-mentalism and logical positivism ultimately proved problematic as a foundation for the field (see Tomalin 2006 for a history).

Devitt (2006) offers a philosophically nominalist position as an alternative to both Chomskyan mentalism and Platonism as envisioned by Katz and Postal. He argues that linguistics is an empirical science which studies languages as they are spoken by linguistic communities and viewing sentences as *abstracta* or rather 'idealised tokens' is nothing more than a theoretical convenience. Devitt's 'linguistic view' (as opposed to the 'psychological view') claims that grammars map onto behavioural output of language production, of which speakers are generally ignorant. His "first major conclusion" is that "a grammar is about linguistic reality not the language faculty. Linguistics is not part of psychology" (Devitt 2006: 9). He arrives at this and related goals by questioning the *representationalism*, or the idea that speakers tacitly represent the rules of their language, at the heart of the generative programme. We'll return to some of these themes in Chapter 9.

One of Devitt's favourite examples is that of von Frisch's theory of the 'waggle dances' of bees. He uses the theory to make three general distinctions (in the spirit of realism). Von Fisch observed that bees use a form of communication called a 'waggle dance' to indicate the direction and distance of food sources to other bees in the hive. For example, if a bee returns from a food source over 100 metres away, it will employ a waggle dance (a 'round dance' if less than 100). The angle at which the bee arrives in the hive reflects the angle with relation to the sun of the bee's path from the food source while distance is indicated by the speed of the dance. He specifically uses this example (and others) to distinguish between (1) the theory of the waggle dance (a snapshot of which I provided above), i.e., the behavioural outputs of the bee, and the theory of the bees' competence in its execution. Von Fisch's theory clearly only provides insight into the former. Another distinction is between (2) the structure rules of the dance, which can be diagrammatically presented, and the structure of the processing rules of the individual bees themselves (i.e., what's going on when they compute various distal and directional parameters for communication) of which we have no conception. Last is (3), the respect constraint or the claim that "the bee's state of competence, and the embodied processing rules that constitute it, must 'respect' the structure rules of the dance in that they are apt to produce dances that are governed by those rules" (Devitt 2008: 205). It seems clear that von Frisch's theory of bee dances, grammar of their language if you will, is concerned with the structures of the dance itself as per (1) and not the structures of their competence or performance of it ((2)) of which we know nothing except that it respects the rules of the theory in the sense of (3) (his notion of 'respect' is a technical one).

From the above distinctions, Devitt claims that grammars of linguistics are true of an extra-mental linguistic reality and not human psychology (where English, French, or isiZulu are our waggle dances). From this conception of grammars he defines his minimal position (M) below.

A competence in a language, and the processing rules that govern its exercise, respect the structure rules of the language: the processing rules of language comprehension take sentences of the language as inputs; the processing rules of language production yield sentences of the language as outputs. (Devitt 2006: 57)

From the above comments, it is not easy to discern whether the linguistic view of Devitt is an object-oriented theory or not. Nevertheless, there are some clues, not only in his rejection of the views we will discuss in the next section but also in his positive account of linguistic theory as a theory of outputs. These outputs are physical or observable expressions viewed as individual entities. For instance, he states that "the theory of [outputs] is as much concerned with real-world objects as the theories of horseshoes, chess moves, bee dances, and *wff*'s" (Devitt 2006: 26).

In the preface to *Ignorance of Language*, Devitt describes both his initial fascination with and initial resistance to linguistics. He states (of his thoughts during his graduate years) "[s]urely, I thought, the grammar is describing the syntactic properties of (idealized) linguistic expressions, certain sounds in the air, inscriptions on paper, and the like [...] It rather looked to me as if linguists were conflating a theory of language with a theory of linguistic competence" (2006: v). This thought is apparently the seed out of which the main ideas of the book grew. Now most Platonists would agree on the last statement, in fact Katz (1981, 1984) and Postal (2003) stress the alleged fallacy of conflating the knowledge of language with language itself, present in generative linguistics. It is the first claim, that grammars are about 'sounds in the air' and 'inscriptions on paper', that seems to be at odds with the concept of P-languages but not with object-oriented accounts generally. Devitt writes "according to my 'linguistic conception' a grammar explains the nature of linguistic expressions. These expressions are concrete entities external to the mind, exemplified by the very words on this page" (2008: 249). Again, if languages are composed of these concrete entities, then language is defined in an object-oriented framework since concrete entities inherit ontological independence from other concrete ordinary objects with which they are identified.

The primary objections to Devitt's proposal have come in terms of its naturalistic reach. In fact, his critics argue to a certain extent that he has 'domesticated' linguistics in the sense of 'domesticated science' (Ladyman and Ross 2007). For instance, Ludlow (2009) claims that "while Devitt purports to be offering a proposal that is faithful to linguistic practice, the range of linguistic phenomena and explanation he surveys is limited" (394). This limitation cannot, for instance, deal with postulates of covert material in syntax (which have no phonological expression), such as PRO which is the assumed subject of infinitival clauses (also see Collins 2007, 2008a). If our structure rules concern physical tokens (sounds, writings, etc.) then elements which do not overtly appear through these media pose a problem. Much of linguistic practice and methodology involves the use of assumed entities or items. Katz (1971) linked the Chomskyan revolution in linguistics to the Democritean revolution in early scientific thought in that it aimed to expose the underlying reality behind appearances. Of course, Democritus also left behind a metaphysical legacy of atomistic thinking which might be incompatible with contemporary physics.

Devitt can, and does, respond to these issues, in part relying on the role of conventions (Devitt 2008). But both the criticisms and responses rely on various covert material or unobservable postulates of the theory conceived of as individual objects in need of ontologically independent explanation. I think this is the wrong way to think about these issues and unlikely to result in any clear resolutions, as I will argue below.

2.3 State and Network accounts

An alternative to the object-oriented views above are views on the ontology of language which draw from models of biological organs, organisms, or populations of individual entities. Again, some views can be considered nominalist while others seem more metaphysically committing according to traditional conceptions. What interests me about these proposals is their focus on categories beyond individual entities. In this section, we discuss the Chomskyan thesis of language as a state of the mind/brain and inferentialist and socio-linguistic conceptions of language as a network of norms and dispositions in use. The first is thoroughly internalist while the second can be considered externalist.

2.3.1 Chomsky and *I-languages*

In Chapter 1, I briefly outlined some of the core tenets of generative linguistics. In this section, I'll focus my attention on the ontology that accompanies the theory.

This is in many ways the dominant ontological position on language which owes that status to the success of generative grammar in the mid to late twentieth century. Although Chomsky's (1957) *Syntactic Structures* set the stage methodologically for research on the structure of natural languages, it was the later *Aspects of the Theory of Syntax* (1965) which firmly established the ontological agenda. In this work, linguistics was deemed a subfield of psychology in that linguists were discovering aspects of the state of language users' mind (where the term 'mind' was understood to be a different level of talking about the brain). At the time of the classical cognitive scientific revolution of the 1950s, language proved an essential platform for the emerging study of the mind. The experimental clarity of behaviourism made it highly influential in North American psychology and the strictures of logical positivism reigned in any ontological expansions beyond a parochial conception of parsimony. Again, we'll return to some these issues in Chapter 9.

Hence, the commitment to a mentalistic ontology of linguistics marked a revolutionary point in the history of the study of the mind. The study of languages, therefore, acts as a conduit for the study of 'the mental'. There are many moving parts to this analysis, so I beg the reader bear with me in my reconstruction. Chomsky describes the precise position in the following way:

> We can take a language to be nothing other than a state of the language faculty [...] So let's take a language to be (say, Hindi or English or Swahili) a particular state attained by the language faculty. And to say that somebody knows a language, or has a language, is simply to say their language faculty is in that state (2000b: 8).

There are a couple of terms to unpack here. Besides what is meant by philosophical terms such as 'state' and 'knows', technical vocabulary like 'language faculty' also need defining. The idea is that to know English is to be in the English-brainstate (appropriately idealised, of course). 'Know' here is divorced from its intuitive or commonsense meaning. In fact, after some philosophical uproar, the term was changed to 'cognise' in Chomsky (1980b). While keeping the propositional character of *knowledge-that* or as Chomsky states the term "has the structure and character of knowledge", it differs in the applicability of concepts such as 'justification' or 'warrant' usually associated with the philosophical literature. In fact, for Chomsky 'cognise' marks a tacit form of knowledge of the rules of language, one which is often consciously inaccessible to the non-linguist (and even the linguist who explicitly knows the rules). These rules are represented by a generative grammar or computational procedure. Chomsky (1991: 9) claims that "there can be little doubt that knowledge in a language involves internal representation of a generative procedure".

I'll take what is meant by 'state' and 'language faculty' together here as they are interrelated. The language faculty is a biologically endowed computational system responsible for the generation of language (or more specifically 'narrow syntax' in later versions of generative grammar). In other words, it is a system of rules embodied or supervenient on brain-structures of language users. This system is in turn characterisable in terms of states.

Linguistics is simply that part of psychology that is concerned with one specific class of steady states, the cognitive structures that are employed in speaking and understanding. (Chomsky 1975: 160)

These 'steady states' correspond to individual languages and they are produced by a generative procedure or set of recursive rules known as a generative grammar. Here Chomsky identifies both the subject matter of linguistics and the generative procedure with an I-language or "[o]ne component of the language faculty is a generative procedure (an *I-language*, henceforth language) that generates *structural descriptions*" (Chomsky 1993: 1). Think of structural descriptions as linguistic trees for now. The important point is that he thinks "the state attained is a computational (generative) system" which "[w]e may call that state a language or [...] an I-language" (Chomsky 2000b: 78).

Thus, the study of language became the study of *I-language* or an internalised system of generative rules for the construction of steady states corresponding (roughly) to natural languages like English. Internalisation is only one of the senses supposed to be evoked by the 'I' in I-language. There are three other senses in the concept. Ludlow outlines them in this way:

First, [Chomsky] is addressing the fact that we are interested in language faculty as a function in *intension* - we aren't interested in the set of expressions that are determined by the language faculty (an infinite set we could never grasp), but the specific function (in effect, grammar) that determines that set of expressions. Second, the theory is more interested in *idiolects* - parametric states of the language faculty can vary from individual to individual (indeed, time slices of each individual). Thirdly, Chomsky thinks of the language faculty as being *individualistic* - that is, as having to do with properties of human beings that do not depend upon relations to other objects, but rather properties that supervene on events that are circumscribed by what transpires within the head of a human agent (or at least within the skin). (2011: 47)

Ludlow thinks that the last of these is the weakest link and thus not required by the overall theory. He moves for a concept of a 'Ψ-language' which allows for the possibility of the I-language to relate to external objects either constitutively or otherwise. This amendment would make possible a marriage between the computationalism of generative grammar and externalist theories of meaning (Kripkean, Burgean, or otherwise). For my purposes, it also allows a more inclusive marriage between contemporary non-cognitivist accounts of cognition in terms of the 4E approaches (see Chapter 9 for more details).

Returning to the issue of ontological commitment, we can see quite clearly that this view of language is not subject to object-orientation or ontological independence. The internalised system of linguistic rules common to all human beings, or UG where "UG may be regarded as a characterization of the genetically determined language faculty" (Chomsky 1986: 3), is like a network of connected nodes. The famed Principles and Parameters (P&P) model of language serves as a good illustration of this alternative metaphysical picture.

The basic picture evokes the image of a switchboard. P&P starts with the assumption that we are biologically endowed with a language faculty (as described above). This faculty is prewired with two kinds of linguistic features: principles and parametrized principles or parameters for short. The principles are all fixed and universal whereas the parameters are like binary switches set at either 'on' or 'off'. The initial state of the system or UG sets all the parameters to 'off'. This is the state into which we are born as children. Then, as we are exposed to language in our environments, the parameters of the language faculty begin to be 'set' to either of the binary options. Once all of the parameters have been set (turned 'on' or remain 'off') the faculty of language reaches the steady states we mentioned above. Language acquisition is complete (usually around puberty) and linguistic competence is established. We *know* or *have* our native languages.

An example might help. Some languages require subjects or rather have obligatory subjects. English is such a language. So, even when no subject is present as in the case of *It is raining*, a mandatory pronoun is inserted into the structure. Other languages, so-called 'pro-drop' languages, have no such requirement. Spanish and Russian are such languages. So we can imagine a parameter which specifies whether you are acquiring a pro-drop language or not. If, in your linguistic stimuli, you find instances of either of these options, the switch is set or left off respectively (if you find both, the parameter becomes optional).

There are a host of theoretical problems with the P&P model of acquisition and the language faculty (for a historical overview, see Newmeyer 1996). That is not our direct focus right now though. Rather we are interested in what the image of language is according to such a model. Firstly, the framework is rather structural. A language is a system of structural units approaching a steady state in which those units are relatively fixed. Languages are thus not objects akin to physical or macroscopic objects like tables and chairs but ontologically interconnected systems of properties and features in which the changes in these structures result in changes in the states that correspond to the languages in question. I think the basic picture leaves out an important aspect of dynamism on which I will follow up in Chapter 5. But for now, we can see that the generative linguistic ontology of the language faculty is far from an object-oriented one.[6]

One aspect that is often neglected by the kind of ontological picture described in this section is the social dimension of language and its use. Chomskyans have traditionally sworn allegiance to a competence model as the only scientifically viable option, leaving the performance level relatively underexplored. In the next section, we look at views which attempt to reverse the focus of the science towards a social ontology of language.

2.3.2 Inferentialism and Sociolinguistics

We've seen what an account of language looks like when it eschews commitment to languages as individual objects above. The Chomskyan picture is utterly internalist. The rules of grammar, the I-language, is an internal generative system. Chomskyans are interested in linguistic competence in terms of these internal procedures and they identify language itself as a system of these procedures aiming at a steady acquired state. In Quinian terminology, the theory quantifies over highly interconnected states *not* singular objects. However, in their rejection of E-languages or external languages, they also neglect the social-constitutive aspect of language and its use in actual communities. With Devitt, we saw a view which incorporated linguistic output but maintained an object-oriented individualism. In this section, I want to talk about two views which do something similar without the object-orientation, namely Labovian sociolinguistics and inferentialism respectively.

Both of these views fall under the general rubric of 'public language' accounts. A chief proponent of this concept is Michael Dummett. For Dummett (1993), natural languages are constituted by the social practices in which they are used. Thus, the target of linguistics is the public languages which emerge from shared linguistic activities in a community of language users. These shared practices are sometimes explained in terms of conventions. For instance, Millikan (2005) argues that "a primary function of the human language faculty is to support linguistic conventions, and that these have an essentially communicative function" (25). She thus offers an alternative approach to the Chomskyan view that there is no 'public object' (or E-language) which corresponds to a natural language. However, contrary to many views on the nature of a public language, such as the Platonism of Katz and Postal, Millikan argues that:

> The web of conventions that forms the mass that is public language is not an abstract object but a concrete set of speaker–hearer interactions forming lineages roughly in the biological sense. These lineages and their interactions with one another are worthy of scientific study. Nor are their properties derivative merely from the properties of I-languages. (2005: 28)

These sorts of accounts of linguistic theory emphasise the communicative social and hereditary aspects of natural language. The Chomskyan view, and generative linguistics in general, is more of a *synchronic* theory which looks at a snapshot of the transition from the initial state to mature language in an idealised individual. *Diachronic* accounts go back further since their objects of inquiry are communities of agents not individual language users. In Chapter 5, we will return to the biological analogy but at the level of systems.

A specifically semantic proposal in this vein is contemporary Brandomian inferentialism (Brandom 1994, Peregrin 2015). The core idea of this framework is that in the same way that the meanings or content of logical constants can be determined by the inferential roles they play in a logical system (*via* introduction and elimination rules), the meanings of ordinary terms and words in natural language are determined by their inferential roles. This is a radical idea which challenges a number of tenets of the referential/truth-conditional theories of meaning the likes of which have dominated the semantic landscape since the work of Frege and Russell in the early twentieth century. I'll take up some of these arguments in Chapter 8 again.

In a recent exposition of the inferentialist programme, Peregrin (2015) argues for the central place of rules within linguistic semantics. Unlike the Chomskyan conception of rules which are internal and individualistic (two aspects of the "I" in I-language) and the external communicative aspects of natural language are purely ancillary in most respects, rules are norms established by the practices of linguistic communities.

> To be sure, if an expression has a meaning within a linguistic community, then the speakers of the community *will* conceive of it in certain specific ways. However, this is not enough to establish the fact that it means what it does. An essentially private act of conception is not capable of grounding the essentially public institution of language. That people of some community mentally associate the word *spider* with a certain kind of animal is a fact

of their individual psychologies not capable of establishing the fact that *spider* expresses, within their language, the concept of *spider*; what is needed alongside any private associations are some public practices that make the link between the word and a concept public and shared. (Peregrin 2015: 44)

Although the concept of rule remains central on this view, the order of explanation is reversed from the inside-out to a view of meaning being determined from the outside-in, from external social normative practices to internal mental representations. Thus the target of linguistics is then more sociological than psychological.

Importantly, individual languages on this view are not social objects but rather networks of interconnected behavioural regularities governed by norms. To speak English or to know English, is to be embedded in the norms and practices of English speaking communities. In this sense, it's like a language game in the Wittgenstinian sense where this involves the claim that "[e]very language is constituted by a language game, that is, a set of rules for the use of words together with related non-linguistic action" (Priest 2006: 190). Another way of putting this point is that Chomskyans take languages to be systems of internal rules and inferentialists take languages to be systems of external rules. Of course, the latter philosophers also described these rules in explicitly normative terms whereas many of the former would disavow such talk.[7]

The empirical linguistic project that best dovetails with an inferentialist persuasion can be found in Labovian sociolinguistics. This research programme takes language acquisition and language change to be central *explananda*. As Labov states of the former,

[W]e are programmed to learn to speak in ways that fit the general pattern of our community. What I, as a language learner, want to learn is not 'my English' or even 'your English' but the English language in general. (2012a: 6)

That there is a viable scientific object of study in 'the English language in general' has been a matter of some dispute, especially from Chomskyans. But what Labov showed is that there are complex regularities of languages discoverable only at the social scientific level. For instance, "[Labov] soon established that when the collective pronunciation of a certain vowel changes without being noticed by the speech community, it is likely that upper-working class or lower-middle class women are leading the change" (Woschitz 2020: 163). He further showed that in various societies, the lower-middle class tends to accommodate upper-class pronunciation, another process led by women.

Such observations formed the development of 'language change principles' not unlike the principles of language variation we saw in the P&P model earlier (Labov 2001, 2010). Except, here the parameters are gender, class, socio-economic status and so on.

This framework leads to a view of the ontology of language and its scientific study at the level of social facts. Labov himself admits that "[t]his work strongly supports the central dogma of sociolinguistics that the community is conceptually and analytically prior to the individual [...] language is seen as an abstract pattern located in the speech community and exterior to the individual" (2012: 266). So, Labov considers languages to be 'abstract patterns' located in speech communities. Again this invites a structural network notion of language, where changes in certain nodes or their relations can have global (or societal) level effects.

What is interesting from our perspective so far is that both prominent scientific strategies for grounding the study of language, generative grammar and Labovian sociolinguistics, although different to the point of oppositional, agree on the rejection of object-orientation. In the pages to come, I will argue that this is no accident as linguistics—across frameworks—is an essentially structural enterprise.

3

The Many and the None

In this chapter, I want to briefly consider two last alternatives to the views of
the previous one, both left out by the Katzian trichotomy influenced, as it was,
by a priori metaphysical categories. We'll start with anti-realist accounts and
then move on to pluralist alternatives. The central insight of this chapter will
be along the lines of the following statement:

> CENTRAL INSIGHT II: ALTHOUGH ANTI-REALISM AND PLURALISM BOTH
> POINT TO IMPORTANT EXPLANANDA IN THE ONTOLOGY OF LANGUAGE, THEY
> FAIL TO PROVIDE ADEQUATE ACCOUNTS FOR ADDRESSING THEM.

3.1 Anti-realist Accounts

It is important at the outset to distinguish between different kinds of anti-realist
theses in the philosophy of linguistics. There are what I call the 'global anti-
realists' who think that language itself does not exist, at least not in the ways we
usually think it does. Then there are those we will deem 'local anti-realists' who
argue that individual linguistic items or standard linguistics entities (SLEs) do
not exist in any robust or scientifically realist sense. These views are best kept
distinct, since one could be a global anti-realist without adhering to the local
variant and vice versa.

Both kinds of the accounts discussed in the previous sections are committed
to a certain kind of scientific realism about the objects, states or networks of
natural languages and linguistic theory. I'll pick out three candidate anti-realist
views in this section. We'll start with the local option.

3.1.1 Rey on *Intentional Inexistence*

Georges Rey (2006, 2020) begins his account with definitions of 'standard
linguistic entities' or SLEs as tokens of word, sentence, morpheme, and
phoneme types and 'physical tokenism' or PT as the assumption that SLEs can
be identified with physical (*acoustic*) spatio-temporal phenomena. The aim

Language, Science, and Structure. Ryan M. Nefdt, Oxford University Press.
© Oxford University Press 2023.
DOI: 10.1093/oso/9780197653098.003.0003

of his view is to suggest that SLEs do not share in the objective reality of entities like cars (especially his own reliable Honda). He starts by drawing a distinction between two kinds of representation, *existential* and *purely intentional*. Compare the following two statements:

1. The word "Bush" represents Bush.
2. The word "Zeus" represents Zeus.

In the first instance, a speaker is attempting to refer to an actual person (the former president of the United States). In the second, he claims "when we say things like (2), where we know we aren't referring to any actual thing, we resort to the intentional use of 'represent', shifting our concern from any actual referent to the mere idea of it" (Rey 2006: 242). Essentially, he is using the old problem of empty names in the philosophy of language to generate a technical concept of representation which does not require the existence of the thing represented. So Rey says, "an (intentional) content is *whatever we understand x to be when we use the idiom "represent(ation of) x" but there is no real x*" (2006: 242).

His aim is to hold on to representationalism without giving into idealistic interpretations such that ideas act as referents but only exist in the mind and not in physical reality, etc. How he achieves this task is by claiming linguistic theory is only committed to the intentional contents of SLEs like nouns, verbs, noun phrases, and so on, not actual nouns and verbs in the external world. Thus, purely intentional uses of 'represents' denote 'intentional inexistents' (a term he borrows from Brentano) which are essentially fictions.[1]

Rey then proceeds by asking whether or not the reality of these entities is required by contemporary linguistic theory. He answers negatively. The structural descriptions, e.g., tree diagrams, we associate with sentences do not map onto any real entities. Unlike the structures we associate with objects like cars, sentences and words have no referents in reality. Or as he puts it,

> Whereas examining the spatiotemporal region of my car clearly reveals the existence of the intended structure, examining the spatiotemporal region of my utterance reveals nothing remotely like the linguistic structure that I presumably intended. (Rey 2006: 246)

As evidence for such a position, he cites the well-known fact that there are no natural word or sentence boundaries in acoustic signals sent from our articulatory systems. Specifically, there is nothing in them that would cause "the physical parts of words, let alone NPs, VPs and the like" (Rey 2006: 246). Rey's claims here are of course independently backed up by data generated from sound spectographs and the like (and analogous visual tools should be appropriate for the sign language case).

This would militate against both Platonism and Devitt's linguistic conception. It also goes against the social ontological views of the previous section. Since "if linguistic entities are intentional inexistents, then they are found purely in the realm of the mental, which might tell against the social ontology of language in favor of a psychological one" (Santana 2016: 513). This might indicate allegiance to the Chomskyan picture of language as a mental state or system. But Rey does not endorse that view either. At least not as a location of the objective reality of SLEs. For this conclusion, he uses Kanizsa figures and optical illusions to suggest that psychological reality, i.e., seeing a triangle where there is none, is not reality *tout court*. Those who object have the burden of identifying the contents of these representations with their causes in the external world, as you would if you said you saw a car or a table, but Chomskyans are internalists and so deny such a language-world correspondence pertains.

Overall, anti-realism about SLEs is a compelling position. However, it's not an unavoidable one. Especially if one realises that Rey is looking for reality in the wrong place, or more specifically at the wrong level, namely at the level of objects. In the next chapter, we will offer an alternative ontological picture which can accommodate the worries Rey presents here more broadly without giving up on realism. In Chapter 6, we will present an account of the ontology of SLEs based on that same picture.

Importantly though, Rey isn't denying that language itself exists. For this we turn to Davidson (1986) and Ludlow (2014) for different but related versions of the global anti-realist perspective on language.

3.1.2 A Nice Derangement of Dynamic Lexica

Donald Davidson is one of the most celebrated systematic philosophers of the modern era. His work reaches far into the philosophies of mind, language, and epistemology. In the philosophy of language, he is renowned for reversing Tarski's direction of explanation for truth as relying on a fixed theory of meaning by using a fixed theory of truth to establish a systematic theory of meaning (Davidson 1965). Given these achievements, it might come as a surprise that he later advocated a global eliminativism about natural language.

In his Mrs Malaprop inspired 'A Nice Derangement of Epitaphs', Davidson makes the following claim:

> I conclude that there is no such thing as a language, not if a language is anything like what many philosophers and linguists have supposed. There is therefore no such thing to be learned, mastered, or born with. We must give up the idea of a clearly defined shared structure which language-users acquire and then apply to cases. (1986: 446)

He establishes this conclusion by introducing a puzzle about communication. The idea is something like this: there are two different kinds of theory a given hearer could use to interpret a given speaker's utterance, a *prior theory* and a *passing theory*. The prior theory is of the systematic nature of Chomskyan linguistics or Davidson's own work on semantic theory. It assigns meanings to forms in a systematic manner. But when we actually encounter each other in conversation, we don't seem to use such mechanisms. This is a point contemporary generative and non-generative linguists have been pushing for decades, ideal competence does not give us much traction on performance. When we interpret each other on the fly, we use a passing theory tailored to the situation and idiolects of our interlocutors. The extreme cases of which involve malapropisms or the misuse of words based on their pronunciation, as mentioned by Davidson. But we can also understand second-language users who make syntactic mistakes as further evidence of the phenomena supported by the passing theory thesis.

Davidson's point is a conditional claim and many commentators (or rather defenders) have insisted that his view amounts to little more than a rejection of the idea that formal linguistic theory guarantees real-time interpretative success in conversational contexts. However, I follow Stainton (2016), in rising above the exegetical issues to the question of whether Davidson's argument provides us with enough resources for a global eliminativist position. Seen in this light, Davidson's argument says that the existence of a public language like Mandarin or Cantonese is otiose or theoretically inert since no such entity plays a role in explaining our everyday communication. As Stainton states, "[i]f knowledge of a public language is neither necessary nor sufficient for successful conversational interaction, then public languages are explanatorily otiose" (2016: 9). This, of course, is a strong constraint on explanation and unlikely to be true. Public language might not be necessary or sufficient for a full explanation of conversational dynamics but it might (and probably is) still part of the story.

Ludlow's (2014) more recent account employs a similar eliminative strategy based on how humans communicate. He begins with a familiar property from the radical contextualist literature in the philosophy of language, namely the "the extreme context sensitivity of language". He forcefully jettisons the idea that languages are stable objects that we somehow learn and use, similarly to Davidson: "According to Ludlow, human languages are things that we build on a conversation-by-conversation basis-he calls them 'microlanguages'" (Cappelen 2018: 163). Ludlow uses this claim to argue that word meanings are dynamic and vastly underdetermined, hence 'the dynamic lexicon'. During and within conversations, the meanings of words are under constant change and semantic shift. The passing theory Ludlow endorses is called 'meaning control':

> The doctrine of Meaning Control says that we (and our conversational partners) in principle have control over what our words mean [...] If our conversational partners are willing to go with us, we can modulate word meanings as we see fit. (2014: 83)

There are a number of issues with both Davidson's and Ludlow's proposals. Firstly, I think the level to which we reconstruct our languages on the fly is greatly overestimated by both authors. From an empiricist perspective, each conversation we enter into from our early linguistic interactions are lessons and tools for the creation of our mature language competence. Naturally, there is some interpretative skill and innovation involved. But those things should not be exaggerated. It's up to the science of language to make sense of this complex mixture of formal linguistic structure and real-time processing techniques. In addition, there would need to be an explanation at the level of social facts as well. Nevertheless, a more clear account of structural continuity would assist in preventing these kinds of exaggerations. I hope to provide such an account in the next chapter.

The main issue with both Davidson and Ludlow is that they make a mystery of communication. Human communication is a datum, an *explanandum* even, but not a mystery. Individual conversational contexts might create passing or microlanguages but they can still subscribe to the overarching established rules of a language. Without this fact, explaining how individuals coordinate from one conversational context to the next would be a scientific conundrum. *Contra* Davidson there is shared structure. The aim of the positive parts of this book is to make that claim perspicuous.

If eliminativism or antirealism is not a viable option, then what about its conceptual antithesis: pluralism? It is on to this question that we must now move before concluding.

3.2 Why I Am Not a Pluralist

The previous sets of views offer a sparse ontology either by aiming for empirical adequacy or highlighting paradoxes within the conventional accounts. Pluralist views jettison ontological parsimony for the twin goals of explanatory coverage and unification of methodology. I will review two sophisticated pluralist accounts in this section. The first advocates for a union of methodological approaches but leaves one asking for their ontological correspondence. The second starts with a similar methodological liberalism but ambitiously attempts an intersection of the respective ontologies.

3.2.1 Santana's Union

Carlos Santana (2016) has recently argued in favour of a pluralist ontology for natural language based on all of the major foundational approaches we discussed in this chapter so far. He, like me, isn't interested in the folk concept of language but rather the ontological question through the lens of "what sort of roles the concept of language plays in linguistic theory and practice" (Santana 2016: 501). So far so good. He makes his case by showing that each candidate view—psychological, social, and abstract—provides good positive reasons for its chosen ontology and bad negative ones for the rejection of its rival. This, he argues, naturally leads to pluralism, but not without qualification.

The first thing Santana does is to separate the discussion into two related questions, one scientific and the other metascientific or 'descriptive' and 'normative' in his terms. He claims that "[l]anguage, the scientific concept, is thus descriptively whatever it is that linguists take as their primary object of study, and normatively whatever it is they should be studying" (Santana 2016: 501). His argumentative strategy for establishing his pluralist ontology is an interesting one. He starts by evaluating the positive cases for each ontological position and then dismissing each negative case for the respective ontologies. In this way, the descriptive question answers the normative one. By considering what linguists and philosophers of language actually take natural language to be, we can conclude that we *should* unify the approaches. In other words, if there are no good reasons for rejecting a candidate ontology then we should admit it into our scientific concept of a language.

In terms of naturalism, which will be covered more thoroughly in the next chapter, Santana constrains both his argument at the onset and his conclusions on scientific grounds. He lists three criteria for determining the proper subject matter of linguistics, "(1) x exists, in a form accessible to scientific study, (2) x is (descriptively) a primary object of study for linguists, and (3) x is reasonably referred to as 'language'" (Santana 2016: 503). Then, he insists that pluralism entails the rejection of agnosticism in terms of ontology. And moreover, if this is the case, then methodological isolationism would be a scientifically unwise trajectory on which to continue.

Besides the entire discussion being laden with object talk, which we can allow for the moment, there is one serious omission in Santana's account. The problem I have is with the Venn diagram of his argumentative strategy. He makes two telling moves, one to establish that each ontology has a piece of the puzzle of language to contribute, then another to show (effectively, in my view) that the usual reasons for rejecting any of them are not very good ones. But the missing link is a story of how these puzzle pieces fit together exactly— and here Santana offers us little guidance. In other words, we have a convincing argument for an ontological union but no idea of how the elements intersect. Furthermore, since it's not obvious how social, psychological, and abstract ontologies are directly compatible in terms of a coherent scientific target, we need more than just an argument stating that they *should* to be. For this, we turn to Robert Stainton's (2014) pluralist account.

3.2.2 Stainton's Intersection

Stainton (2014), like Santana, aims to incorporate aspects of Platonism, Chomskyan mentalism, and physicalism (i.e., Rey's PT) into one coherent ontology. He does not afford as much attention to socio-linguistics or pragmatics. Nevertheless, he makes a strong case drawing from the heterogeneity of methdology to the hybridity of ontology. I will provide a brief overview and then discuss why I do not believe that this alternative gets to the heart of the issues which form the target of the present work, despite its improvement upon Santana's account.

Firstly, pluralism rejects the idea that there is any real distinction to be had with the concept of 'linguistics proper' assumed in much of the foundations debate (see Katz and Postal 1991), which was coined to isolate syntax and semantics as the only properly linguistic domains.[2] From the pluralist perspective, phonology and phonetics (and presumably pragmatics) are equally linguistic domains worthy of inclusion within any debate concerning the foundations of the subject. Naturally, once we move towards a broader methodological base, the ontological plurality seems to follow. In the spirit of this diversification of the properly linguistic, Stainton describes the corresponding metaphysical attitude in the following way:

> My own view [...] is that natural languages, the subject matter of linguistics, have, by equal measures, concrete, physical, mental, abstract, and social facets. The same holds for words and sentences. They are metaphysical hybrids. (2014: 5)

Stainton provides various arguments based on some extremely interesting evidence to support this ontological attitude. I divine two main lines of argument for the pluralist claim. The first is that the ontological attitudes of the previous sections (with the inclusion of public language views and naive physicalism of the pre-Chomskyans) all have something to contribute to the subject matter of linguistics and therefore its corollary that each ontology misses out on something essential about the overall picture individually. Here, his view dovetails with Santana's. The second line is that on a specific reading of these ontological commitments, and contrary to *prima facie* impressions, these ontologies are perfectly compatible with one another. This was the missing link in the pluralist Venn diagram, I mentioned above.

Let us consider some of the evidence for the first claim, i.e., that physicalism, mentalism, Platonism, etc. all have some important *sine qua non* piece of the linguistic puzzle to contribute. I'll briefly go over the argument for the necessity of the physicalist contribution which comes from phonetics. The argument from phonetics simply involves the truism that vocal and auditory organs, and the sounds which they produce, play a role in language production and comprehension. Phonetics concerns the physical movement of the vocal tract, the tongue, aspiration, etc. These phenomena are clear candidates for physical aspects of natural language.

The cases for the inclusion of the mentalism and Platonist ontologies have already been covered above in Sections 2.2.1 and 2.3.1. The point is that all of these facets contribute to the 'properly linguistic' and thus the exclusion of any of them is tantamount to incomplete characterisation.[3]

Despite the ecumenical spirit of the approach, there are a few aspects of this line of reasoning which I take to be questionable. Firstly, there is a distinction between the fact that a contribution is made by a set of phenomena to the description of a general (super)phenomenon and whether or not it counts as detrimental to abstract away from it. The idea of 'linguistics proper' is an abstraction in my view. In the process of scientific investigation abstraction is a necessary tool. In many cases this process involves the omission of potentially connected or relevant material (see Weisberg 2007). For example, the way Stainton frequently mentions that a certain phenomenon is not 'exhausted' by a certain description or methodological characterisation is misleading as scientific investigation might not be interested in 'exhausting' descriptions of phenomena but rather *minimally* representing them for different explanatory or predictive purposes.

Consider the case of phonetics. It is undeniable that most of the world's languages involve sounds in terms of combinations of vowels and consonants. Some of these combinations are quite complex, such as the click sounds of Nguni languages of my home country of South Africa.[4] Nevertheless, they are not essential to an understanding of natural language *simpliciter* since there are languages which do not produce sounds at all. Sign languages operate on manual communication with physical gestures and signals to convey information. Some sign languages include haptic cues (used in communities whose members suffer from both deafness and blindness). They have syntax and semantics. They do not have phonetics or phonology, in the traditional sense at least. Assuming that phonetics contributes some essential aspect of natural language without which we do not have a characterisation of natural language relegates sign languages to the camp of the non-natural. Abstracting away from phonetics could therefore be considered scientifically benign or even necessary for deeper understanding.

Let the above objection serve only to question the pluralist agenda. My purpose is not to produce knockdown arguments here but rather to suggest how a scientist might abstract away from some linguistic material and arrive at a foundational project involving only syntax and semantics as 'linguistics proper'.[5] Consider Newton's theory of the tides. The earth's rotation and the gravitational pull of planets besides the earth, sun, and moon are abstracted over. These factors can or do play a role in tidal force (the ancillary force of gravity responsible for the tides) but they are not part of the core explanation of the phenomenon.

Moving on to the second line of argumentation for this view, let us consider the metaphysical pluralist claim. This argument is a response to an immediate objection along the lines of Postal (2003, 2009) as to the incompatibility of the various ontologies associated with mentalism, Platonism, physicalism, and public language views. Stainton begins the pluralist apology in this way,

There is an obvious rebuttal on behalf of pluralism, namely that "the linguistic" is a complex phenomenon with parts that belong to distinct ontological categories. This shouldn't surprise, since even "the mathematical" is like this: Two wholly physical dogs plus two other wholly physical dogs yields four dogs; there certainly is the mental operation of multiplying 26 by 84, the mental state of thinking about the square root of 7, and so on. (2014: 5)

However, similarly to the result of thinking about the square root of 7, I do not believe this is analogy is rational. Mathematical reasoning does indeed involve mental operations, some physical examples (instantiations?) and the like but this is not usually how people conceive of 'the mathematical'.[6] On the standard picture, mathematics is considered an abstract science. Mathematicians study mathematical objects and rules which often outstrip the physical and the mental. The processes involved in mathematical thinking are certainly within the realm of psychology, but the fact that physical objects obey rules of arithmetic is not enough to hold arithmetic to purely physical characterisation. Does the question of how many dogs are in the union of an infinite set of dogs and another infinite set of dogs receive a physical interpretation? Stainton does acknowledge that natural languages display 'interdependence' of these factors which perhaps 'the mathematical' does not.

His main argument against incompatibility, and in favour of intersection, is that the former rests on an equivocation of the terms 'mental', 'abstract', and even 'physical'. Once the equivocation is cleared up, it is argued, hybrid ontological objects are licensed. Let us briefly consider the different senses of these words proposed by Stainton. Physical$_1$ is something like an object under the purview of the hard sciences such as physics. Weeds, loosely defined as unwanted plants, would not count. On an extensional physical$_2$ definition, weeds show up since they have spatio-temporal and other physical properties. Mental$_1$ includes individual mental states such as pains and hallucinations. Mental$_2$ involves a specialised notion of secondary qualities conditioned by the mental but not identifiable with mental items. Aspects of taste and perception are suggested as examples of mental$_2$. Stainton uses the term 'mentally conditioned' to capture this kind of mind-dependence. Lastly, he contrasts abstract objects *qua* Platonic objects, with what he calls 'abstractish' objects, neither in the mind nor concrete particulars. Musical scores, models of cars and legislation form part of this latter category.

The argument goes that appreciating the physical$_2$, mental$_2$, and abstractish nature of natural language will dissolve worries about ontological inconsistency and open the door for intersection. Consider some other members of this category of objects.

> Indeed, our world is replete with such hybrid objects: psychocultural kinds (e.g. dining room tables, footwear, bonfires, people, sport fishing [...]; intellectual artifacts (college diplomas, drivers' licenses, the Canadian dollar [...]; and institutions (MIT's Department of Linguistics and Philosophy, Disneyworld [...] (Stainton 2014: 6)

I agree that such objects exist in all of the senses which Stainton describes but I think that there is something missing from the picture, something which breaks down the analogy between natural languages and their abstractish cousins. Natural languages are characterised as rule-governed by most linguists. These rules might be inner mental or mental$_2$ in nature but they also constitute something more abstract. Stainton agrees to this much, "it is types whose meaning is compositional and systematic" (2014: 4). But types of this kind seem to have more in common with mathematical theorems and objects than they do with ordinary abstractish ones. This is not the place to go into detail (we'll introduce better vocabulary in Chapter 4). Suffice to say that one feature of the former and not the latter objects is that there are uninstantiated types. There are sentences which have never been uttered, and thus are not physical$_2$ and not obviously abstractish. It is therefore misleading to define such a broad ontological category and throw linguistic objects in with ordinary (abstractish) objects—in terms of either extensions in space and time or secondary qualities with such extensions—in this way. More needs to be said about metaphysical hybrids and their potential subcategories before making any methodological claims based on them.

Furthermore, once the restricted domain of *linguistics proper* is reinstated, and with it goes methodological pluralism, many of the compelling arguments for the metaphysical variety are diminished (such as those involving phonetics and phonology). There are many insights to be had on the pluralist account, many of which I hope to retain, however I offer a distinct approach to the methodological aspects of linguistics, one which is equally compatible with various views on its ontology.

3.3 No Country for Clear Resolutions

There is a relic of *a priori* metaphysical reasoning that permeates many of the above discussions on the ontology of language, namely the type-token distinction. In this last section, I want to make the case that this non-naturalistic kind of reasoning is unlikely to lead to any resolutions on the fundamental questions of the ontology of language. The type-token distinction itself is a familiar tool from the philosopher's toolkit (much like the use-mention distinction). Consider these oft-cited passages (used in both Katz (1996) and Postal (2009)).

> There will ordinarily be about twenty *thes* on a page, and of course they count as twenty words. In another sense of the word *word*, however, there is but one *the* in the English language; [...] it is impossible that this word should lie visibly on a page or be heard in any voice. (Peirce 1958: 423)

> ES IST DER GEIST DER SICH DEN KÖRPER BAUT: [S]uch is the nine word inscription on a Harvard museum. The count is nine because we count *der* both times; we are counting concrete physical objects, nine in a row. When on the other hand statistics are compiled regarding students' vocabularies, a firm line is drawn at repetitions; no cheating. Such are two contrasting senses in which we use the word *word*. A word in the second sense is not a physical object, not a dribble of ink or an incision in granite, but an abstract object. In the second sense of the word *word* it is not two words *der* that turn up in the inscription, but one word der that gets inscribed twice. Words in the first sense have come to be called tokens; words in the second sense are called types. (Quine 1987: 216–217)

Characterisations of objects such as those presented in the quotations above aim to establish a distinction between abstract and ordinary objects. Once this distinction is in place, there are usually two options considered for describing the relationship between these respective kinds. We could go the traditional Platonist route of removing abstract objects from the causal order by stripping them of physical and temporal parts. This is inimical for a number of reasons, chief among them the conclusion that we are no longer dealing with an empirical science. Another option is adopting a position called 'Naturalised Platonism' (Linsky and Zalta, 1995). This position makes the empiricist claim that properties and sets and other *abstracta* are well-within the causal order and knowable *a posteriori*. We'll return to this option in Chapter 6 when we discuss the ontology of words. In some ways, Quine too falls within this camp by constraining abstract objects through the same principles (such as Ockam's razor) that constrain other theoretical entities. Still we are left in some confusion as to how we come to know these entities in the first place. And moreover, the type-token distinction presupposes an object-oriented account of ontology, i.e., how an abstract type *qua* individual Platonic object instantiates a physical token in terms of ordinary physical objects.

A symptom of this general metaphysical ailment, specific to the philosophy of linguistics, is the persistent claim that languages have a dual nature or live in two worlds. Let us call this the *dual nature thesis* (DNT). This thought is especially pronounced in the Platonist literature where it is presented as a *reductio* for mentalist accounts. As Postal explicitly states (with relation to the Chomskyan claim that natural language is a biological system with discretely infinite output):

> Chomsky takes the putative biological entity to be constructing abstract objects, sets. This is entirely distinct from using some piece of formal science to describe something nonformal. And this conflation by Chomsky of formal and physical objects is one of determinants of the incoherence of his ontological position. (2009: 119)

The primary accusation is that Chomsky is confusing linguistic types for their physical (or mental) tokens. But it is not just opponents who tacitly adhere to this metaphysical distinction, but many generative or biolinguists themselves presuppose a dual character to natural language. In a recent work on the possibility of impossible languages, Moro makes the following distinction:

> [A] human language lives in two different environments: outside our brain and within it. So when we ask whether an impossible language exists we are in fact asking a twofold question: a formal one (concerning rules) and a physical one (concerning matter). (2016: 2)

He does go on to present a unified view of each component to make his case for impossible languages. Nevertheless, the idea that languages are somehow both formal and natural or abstract and physical persists throughout the philosophical reflection on language. Some prominent figures in linguistics have recognised the duality and advocated pursuing a sensible methodological approach until more information on conciliation in provided.

The discipline of generative grammar is founded on a crucial distinction between a language considered as a formally specified set of structurally described strings and a language considered as a behavioral repertoire. A grammar is by definition a membership specification for a language in the former sense. Chomsky has not provided any new definition of the notion "grammar" that severs it from its essential function of specifying the membership of a language. Therefore I continue, as is normal in generative work, to use the theoretical term "language" for the set of strings (or string/structure pairs) defined by a grammar, and not for anything ill-defined such as the set of dispositions toward verbal responses that characterizes a particular language user. (Pullum 1983: 449)

Note that this issue does not seem to be of much concern to the biologist, chemist, or social scientist. Questions of the dual nature of animal species in terms of their abstractness and corporeality do not seem to have preoccupied those working in the philosophy of biology to the same extent that they have those working on the nature of linguistics. In the philosophy of language literature, perhaps no figure exemplifies the two faces of language more apparently than Wittgenstein.

If the young Wittgenstein gave impetus to the development of the calculus conception of language, the later Wittgenstein, like Captain Nolan in the charge of the Light Brigade, tried vainly to rectify the misdirection [...] Language is an anthropological phenomenon. (Hacker 2014: 1272)

Hacker pits the language as a formal meaning calculus view directly against social normative views like inferentialism. In fact, the most prominent philosophical account of language which both submits to the dual nature thesis and attempts a Hegelian synthesis thereof, is given in Lewis (1975). He offers an account of how human beings use languages, construed as abstract objects (or functions from sentences to intensions). For Lewis, languages *are* abstract objects or functions which assign meanings to sets of strings (sentences), i.e., formal objects. Such a language is then utilised by a community of speakers if and only if there is a convention in that community of *truthfulness* and *trust* in that language. Different conventions result in the use of different languages. The definitions of the former terms are of no particular use to us here. Essentially, it's a modelling relation where an abstract object models the behaviour of a system.

What have languages to do with language? What is the connection between what I have called *languages*, functions from strings of sounds or of marks to sets of possible worlds, semantic systems discussed in complete abstraction from human affairs, and what I have called *language*, a form of rational, convention-governed human social activity? We know what to *call* this connection we are after: we can say that a given language L is *used by*, or is a (or the) language of, a given population P. (Lewis 1975: 7)

Lewis' account has been highly influential in the philosophy of language. Again, synthesis admits duality, language bifurcated, even if it is our task to overcome it. As Lewis concludes, "[t]he thesis [that language is an abstract object] and antithesis [that language is a social phenomenon] have been the property of rival schools; but in fact they are complementary essential ingredients in any adequate account either of languages or of language" (1975: 35).

To some DNT is a contradiction or a paradox, to others its a scientific *explanandum*, and to yet others it is an essential union. We have seen all of these accounts represented in this chapter. Indeed, I'm not denying that there is an *explanandum* here in terms of how to relate formal grammars or theories to natural systems or structures. What I am denying is that *a priori* metaphysical reflection based on the traditional type-token distinction will aid us in any way in our explanatory task. Rather the best way of approaching the relationship between the 'formal mode' and the 'material mode' (Ladyman and Ross 2007) is by making use of the resources of the philosophy of science, where these questions have been asked at various levels. The means by which I will pursue this task is by an exploration and redefinition of Dennett's (1991) concept of a 'real pattern'. This is the task of the next section specifically and the rest of the book holistically. Put in stronger terms, the type-token distinction in metaphysics has created more problems than it has solved. In the forthcoming pages, no special status will be given to such metaphysical claims nor the traditional categorisations of linguistic ontology in terms of Platonism, nominalism, and mentalism.

4

Language and Structure

In the previous chapter, I surveyed the landscape on the ontology of language. There was much pruning to be done. I advocated for the rejection, among other things, of object-oriented views based on an unhelpful type-token distinction imported from traditional metaphysics. I left open the possibility—with one prominent brand of modern linguistic theory—that different languages are best identified with systems of interconnected features or properties.

The central insight to be defended and developed in this section is more positive in nature:

> **CENTRAL INSIGHT III**: LANGUAGES ARE NON-REDUNDANT REAL PATTERNS, THE STRUCTURES OF WHICH ARE CAPTURED BY FORMAL GRAMMARS QUA COMPRESSION MODELS.

These is a lot to unpack in this multi-part claim. What are 'real patterns'? What is structure? How are formal grammars compressors? Each topic will receive proper treatment in this chapter. In Sections 4.2–4.3 we'll consider the first three questions and in Chapter 5, I'll suggest a novel interpretation of biolinguistics in order to ground the present analysis. But before then, some further stage dressing is in order. As mentioned throughout the book so far, I see my task as a naturalistic one. However, despite its general popularity and influence, naturalism in philosophy is not entirely easy to define. In the next section, I will describe the general picture along a continuum from extreme naturalism favouring only the posits of fundamental physics (or those in conjunction with other special science posits) to a more moderate variety which is open to resources and claims from a much broader class of 'sciences' but still constrained by physics. My position will fall somewhere in the middle and I'll call it 'moderate naturalism'. My focus will be on linguistics as an especially interesting case of a special science.

Language, Science, and Structure. Ryan M. Nefdt, Oxford University Press.
© Oxford University Press 2023.
DOI: 10.1093/oso/9780197653098.003.0004

4.1 Moderate Naturalism

4.1.1 What is Naturalism?

Naturalism is not a novel position in the philosophy of linguistics. Many theorists have advocated for a 'science first'-approach in their work. Prominent examples are Chomsky (2000a), Scholz et al. (2022), Johnson (2007), and Rey (2020). The most notable figure is of course Noam Chomsky himself. Chomsky (2000) is the naturalistic spiritual brother of Ladyman and Ross (2007), but for the study of language. In that work, Chomsky sets out to decimate the edifice of *a priori* philosophical rumination on language and the role of commonsense folk linguistics in the scientific pursuit of linguistic knowledge. In it, Chomsky claims:

> A naturalistic approach to linguistic and mental aspects of the world seek to construct intelligible explanatory theories, taking as "real" what we are led to posit in this quest, and hoping for eventual unification with the "core" natural sciences: unification, not necessarily reduction. (2000a: 106)

One could attempt to interpret the naturalism advocated by Chomsky here by means of Ladyman and Ross's two core principles for a project in naturalistic metaphysics or what Hendry (2021) calls their 'manifesto for a naturalistic metaphysics'.[1] For them, the only role metaphysics can play is as a grand unifier. This role does not involve the creation of theories or explanations based on intuition or *a priori* logical reasoning nor the disingenuous lip-service to scientific theory (what they call 'the philosophy of A-level chemistry'). Thus, any genuinely useful scientific philosophy or naturalistic metaphysics has to respect two principles, 'the principle of naturalistic closure' (PNC) and 'the primacy of physics constraint' (PPC). The first emphasises the unifying role of philosophy in the following way:

> Any new metaphysical claim that is to be taken seriously at time *t* should be motivated by, and only by, the service it would perform, if true, in showing how two or more specific scientific hypotheses, at least one of which is drawn from fundamental physics, jointly explain more than the sum of what is explained by the two hypotheses taken separately. (Ladyman and Ross 2007: 37)

The principle is subject to a few stipulations on what a scientific hypothesis is, what 'specific' entails and what constitutes a research project. They favour an institutional understanding of science and the scientific method in general whereby "science is [...] demarcated from non-science solely by institutional norms" (Ladyman and Ross 2007: 28). These norms include things like peer-review and regulatory rules about experimentation and theoretical representation in general.[2] And since the practice of science under these rigorous conditions can often result in disciplinary specialisation, metaphysics "is the enterprise of critically elucidating consilience networks across the sciences" (Ladyman and Ross 2007: 28). The PNC is meant to guide and constrain this practice.

How does it achieve this latter goal? One way is by making more precise Chomsky's hope of 'the eventual unification with the "core" natural sciences'. Specifically, the PNC requires that to be considered naturalistic, our endeavour must relate to at least one 'specific scientific hypothesis' of fundamental physics where that term involves a claim which has or is being investigated by institutionally accepted scientific activity (and is fundable by such entities). This is a strong constraint on naturalism. In the literature with has followed Ladyman and Ross, many have relaxed the view to accommodate the *sui generis* postulates of the special sciences.

The requirement on physics forming part of the union of joint explanation of course presupposes the priority of explanations from fundamental physics in any metaphysical task. But since the relation between physics and the special sciences is an asymmetric one, I think we can achieve the spirit of the PNC (and PPC) without embracing a heavy-duty naturalism which would require aspects of formal grammatical theory to find a union with specific hypotheses in fundamental physics. All that we should require is that the former do not contradict the latter posits, and this is a comparatively weak constraint.

Chomsky's own position goes a bit further to insist on what he calls 'methodological naturalism' in which language and the mind are "to be studied by ordinary methods of empirical inquiry" (2000a: 106). Neither the PNC nor the PPC recommends methodological naturalism of this sort. Of course, Chomsky's position is especially curious given his longstanding advocacy against statistical corpus-based approaches to linguistics and the general Chomskyan defence of the role of intuitions as the primary data for theory.[3]

Getting back to naturalism, the PPC does require some interpretation to be of naturalistic guidance for us. Especially if we are to join Chomsky in his skepticism of reductionism. I think one simple heuristic is brought out nicely in a clarifying footnote from Ladyman and Ross (2007): "what we mean is that when a conflict is actually found between what fundamental physics says there isn't and a special science says there is, revisionary work is in order for the special science in question" (190). It is this nuance that I think Fodor (1974) misses on the topic of the relationship between physics and the special sciences. There he introduces what he calls 'token physicalism' or "the claim that all the events that the sciences talk about are physical events" (Fodor 1974: 100).[4] According to him, the special sciences, such as economics, psychology, and linguistics cannot be entirely reduced to physics. Such a reduction would risk their status as autonomous sciences.

However, Fodor argues that special sciences can retain their autonomy, since reduction is possible only with respect to the events they describe ('tokens'), not with respect to properties or natural kinds ('types'). For example, economics describes types of economic transactions such as monetary exchanges which are, of course, physical events albeit heterogeneous ones. These exchanges are multiply realisable from a purely physicalist perspective. In other words, the token events belong to physical explanation but the types do not. Fodor's target is reductions *via* biconditional bridge laws. But such reductions are not the only options to pursue a reductionist project (see Ladyman and Ross, chapter 1, on 'Nagelian reductions'). And moreover, no such project is in fact required by the PPC. What is required is that we draw our metaphysical conclusions from an amalgam of specific hypotheses, at least one of which is based in fundamental physics. And in addition, if the posits of our special sciences contradict posits of fundamental physics, we should adjust the former to resolve the conflict.

I think that the PNC and the PPC so stated are too demanding for a project outside of pure metaphysics or the philosophy of physics.[5] One serious problem with adopting them wholesale is that specific hypotheses of fundamental physics are not settled in their interpretations. Expecting practitioners in the philosophy of the special sciences to undertake such interpretation could, at best, only lead to confusion. Take the claim that ontic structural realism is the correct interpretation of quantum physics as an example. It relies on arguments like French and Redhead's (1988) to the effect that quantum mechanics takes only relations (over individual objects and properties) as ontologically prior. But as Ainsworth points out "there is no interpretation of quantum mechanisms that takes only relations as ontologically primitive" (2010: 53). Whether this claim is true or not is surely not a matter to be settled by philosophers of the special sciences. Perhaps this isn't 'specific' enough? After all, structural realism is a philosophical not a scientific hypothesis (if we allow ourselves the distinction for a moment). Ladyman (2017) provides a nice case study which might illuminate matters. Metaphysicians like to talk about composition. Very often when they do so, they make use of a melánge of logic, mereology, spatial intuitions, and 'lip-service' to science in the form of comments about sub-atomic particles and the like. This can lead to properties such as perceptions of synchronic objecthood taking precedence over dynamic and diachronic analyses of matter over time. In fact, as Ladyman insists, the remit of condensed matter physics is precisely where philosophical questions of composition should start. But I take his general point to be that when philosophers are interested in physical phenomenon like matter and composition, they should consult the science, i.e., physics, that deals with this topic as opposed to their intuitions about ordinary objects. Therefore, similarly when philosophers are interested in linguistic structure and meaning, they should look to the linguistics that studies these aspects of reality before consulting their pre-theoretical intuitions. In other words, they don't need to look at textbooks on the physics of acoustics or to include specific hypotheses from this field in their theorising about language. But of course, neither should they accept any contradictions of that physics in their work.

This might sound like the suggestion that all sciences are equal when it comes to naturalism. But this is far from the case. Ladyman and Ross are surely correct to insist that some sciences are *more equal than others*, i.e. fundamental physics. But acknowledging this fact doesn't seem to lead to anything as strong as the PNC in the form they put it. So why can't naturalism just take the science seriously, whatever science we are reflecting on philosophically? For example, MacArthur (2015: 569) defines both ontological and methodological naturalism in terms of commitment to the 'existence' and 'methods' 'supplied by successful sciences' period. MacArthur (2010) distinguishes between different versions of naturalism in terms of which sciences are sufficient. 'Extreme' admits only physics while 'Broad' includes the social sciences. There are many broad varieties of naturalism on the market. Hutto and Satne work within what they call 'Relaxed Naturalism':

> A relaxed naturalism is one that avails itself of a wide range of scientifically respectable resources, drawing on the findings of a wide range of sciences that includes not just the [natural] sciences but also cognitive archaeology, anthropology, developmental psychology and so on. (2017: 6)

If what is meant by 'not just the natural sciences' is 'not incompatible with fundamental physics' then relaxed naturalism might be PPC-friendly. If the collection of 'respectable resources' can involve posits and claims from various special sciences exclusively then this view captures the essence of my project.

Nevertheless, one might worry that licensing everything that considers itself a science in terms of naturalism is not relaxed but lazy. Practitioners of literary criticism and psychoanalysis might consider themselves to be scientists. And in one short step, we're back to the demarcation problem and the Great Method Debate. 'Respectable' is of no help here either. Ladyman and Ross's institutional gloss also won't protect us from vagaries of funding and special interest groups (see note 2). Similarly I might have been charitable to suggest that 'not just' implied 'not incompatible with' and not some sort of disjoint union of all sciences. So perhaps the best course of action is to think of naturalism in terms of *naturalisation* as a process for a particular target concept or domain. Extreme naturalism á la Ladyman and Ross defines the process in terms of both the PPC and the PNC as we have seen. Lazy naturalism suggests that any claim to science will do. This leaves the possibility of a position somewhere in between (leaning perhaps more towards the extreme than the lazy, I'll admit). So let us define the process in the following way for any posit P or domain D (adapted from Seager 2000).

P or D has been *naturalised* iff:

1. P or D has been explained in terms of some X.
2. X is does not contain P or D as a subset (or does not assume P or D).
3. X is based in the appropriate scientific field(s) related to P or D.
4. X does not contradict the current claims of fundamental physics.

I think this schema captures Ladyman's (2017) example as well as our present subject matter without being too liberal or too conservative. The first two conditions ensure that the subject matter is not presupposed by the process of naturalisation. For instance, if one wants to explain the emergence of content or intentionality, then explaining it in terms of some other process or state X which presupposes intentionality would be problematic (see Seager 2000, and also Davidson 2001). It's important to note that this process doesn't entail ontological reduction. It is epistemic in nature and concerns how we explain a particular phenomenon. Then (3) aligns the target with the appropriate scientific tools for its investigation. In Ladyman's (2017) case, the metaphysics of composition with condensed matter physics. In our case, linguistic structure with syntax and semantics in contemporary linguistic theory. The plural importantly allows for multiple scientific sources to play a role in the naturalisation of a particular target phenomenon. 'Appropriate' is admittedly vague. Perhaps 'evidence' from how ghosts communicate might be relevant for our study of human language. This is in part where condition (4) features, in order to constrain the physical and scientific options by means of an ontology accepted by natural scientists. It applies to the special sciences specifically, since if the appropriate scientific field is physics itself, the condition is redundant.

We'll stay true to this general process in our investigation of what language is and how it relates to linguistic grammars. Before we do so, a note about the special sciences and what makes linguistics a unique member of the class.

4.1.2 A *Very* Special Science

So far we have been assuming an intuitive understanding of what a special science is. But given the general rejection of the role of intuitions in philosophy, it might behoove us to be more precise going forward. In this section, I want to suggest that linguistics is unique among the special sciences in its formalisation and the nature of its explanations. But first, what is a special science?

Again, according to Ladyman and Ross almost everything besides fundamental physics counts as one. So even some parts physics, such as acoustics, are considered special in that sense. Specifically:

> [A] science is special iff it aims at generalizations such that measurements taken only from restricted areas of the universe, and/or at restricted scales are potential sources of confirmation and/or falsification of those generalizations. (2007: 195)

Geology restricts its measurements to the solid surface and sediment of the planet, psychology to aspects of the human mind, and generative linguistics to the subset of those aspects that are exclusively related to language. Of course, to state that of linguistics *simpliciter* is to dismiss the remit of sociolinguistics which is continuous with anthropology and sociology and not only psychology. In fact, one could interpret the debate over the foundations of linguistics in terms of which restricted area of the universe the measurements should be taken from (cf. Santana's (2016) 'normative' question). In this sense, pluralism becomes a surprisingly taxing framework as it requires so many sources of confirmation and falsification without any clear guidelines on how to achieve this task or a ranking between them.

There are other aspects of the special sciences that are often cited in their various characterisations. Kim (1998), for instance, claims that the special sciences mostly explain by reference to causal relations. This is generally not the case in linguistics. Similarly, the quantum entanglement of quantum mechanics which makes individual objects unlikely entities of a realist theory, seem not to be represented in special sciences where "a selfsame rock can freely be transported from context and context without integrity being threatened, and two consumers are necessarily separate if either their utility functions or their consumption histories differ to any degree" (Ladyman and Ross 2007: 196). Again, this is not the case in linguistics where the idea of moving one phrase or expression from one context to another without sacrificing the integrity of the structure is not always viable. Even in sociolinguistics, linguists don't generally assume that the slightest differences in histories or the like would distinguish language users. Language users are idealised in Labovian theory to the extent that "[i]t is argued that the individual does not exist as a unit of linguistic analysis" (Labov 2012: 267). Another claim, this time courtesy of Batterman, to the effect that "the upper level theories of the special sciences are, in general, not yet sufficiently formalized (or formalizable) in an appropriate mathematical language" (2002: 134) is also not a characterisation I would endorse for linguistic theory. As we have seen, linguistics has a historical connection to formal language theory and mathematical logic. Compared to other special sciences, linguistics is highly formalised with many practitioners requiring training in advanced logical and mathematical techniques to model its phenomena. As Johnson states:

> Linguists commonly assume that they are studying fully determinate, math-ematically precise algorithms - *aka* 'grammars' [...] linguistics is distinctive among the sciences in that it does not use mathematical methods to mediate the relationships between evidence and theories. (2007: 396)

Here he is referring to the intuition-gathering practices of generative linguists. But the idea that linguistics lives in two worlds or what we called the DNT in the last chapter is pervasive across the field. Pitt embraces the methodologically heterogeneous nature of the field in his characterisation:

> So far, then, linguistics would seem to be a mixed science, having straight-forwardly empirical departments – orthography, phonetics – whose token objects of study are concrete, and a psychological department – semantics – whose token objects of study are a particular kind of conscious experiences [...] if [...] syntactic structures are literally instantiated by written or spoken sentence tokens, then syntax is a physical science. If they are instantiated

by meanings, then syntax is (on the view developed here) a psychological science. If they are abstract particulars not instantiated by anything, then syntax is a formal science. (2018: 19)

Pitt favours a methodological approach to identifying what a science is, hence the conclusion that linguistics is a mixed science. But his view highlights the formal character of the field. As we will see throughout this chapter, linguistics admits a number of measurements taken from various restricted areas of the universe. In this sense, it is a *very* special science indeed. I believe that this situation has led to the large-scale confusion concerning its foundations we saw in the previous chapter. Thus, approaching the 'specialness' of linguistics from both the naturalised metaphysics perspective and the philosophy of science is the primary goal of the present book. The long-awaited next step is to dissolve the type-token distinction and the DNT with it. For this we need to think of languages is a different way, as real patterns.

4.2 Languages as *Real* Patterns

4.2.1 Dennett's Insight

Dennett (1991) famously introduced the concept of a *real pattern* from complexity and information theory into the service of the philosophy of mind. His target there was a kind of realism about the objects of our intentional behavioural ascriptions (*via* the 'intentional stance'). Specifically, he aimed to show that the success of prediction in folk-psychology is down to its relation to some exploitable pattern in the world. Whether or not he achieved his specific goal is not directly important to our endeavours. What is important is the definition of a real pattern.

The idea is initially quite simple and intuitive. Let us consider a string of a thousand bits of information completely selected at random (you can think of binary code here). In information-theoretic terms, in order to encode a message based on such a string, we would need a computer program at least as long as the string itself. Or as Dennett puts it: "[a] series (of dots or numbers or whatever) is random if and only if the information required to describe (transmit) the series accurately is *incompressible*: nothing shorter than the verbatim bit map will preserve the series" (1991: 32). So the key to identifying patterns is *compressibility*.

For him, patterns are just more efficient ways of representing or describing the initial bit map, i.e. compressions. In algorithmic information theory, Kolmogorov complexity comes closest to expressing these notions. The Kolmogorov complexity of a finite string is defined by the length of the shortest program for a reference Universal Turing Machine (UTM) that outputs that string.[6] In other words, the Kolmogorov Complexity indicates the maximum possible compression of a string and connects this concept to the shortest program that outputs that string.

Dennett's original example involves a series of different patterns (distinct in terms of their random noise ratios), which map onto the same underlying real patterns. But he doesn't distinguish between the sub-patterns and the full pattern they map onto. They are all equally real along a scale of file sizes. The more noise you add, the larger the file size until the pattern is no longer discernible. Essentially what the algorithm does is identify and eliminate redundancies in the string or bitmap. Another way of putting this is that pattern recognition operates by finding regularities in raw data. For Dennett, "compression in general, depends on regularities in human behaviour and, indirectly, on the cognitive processes that generate it" (Millhouse 2021a: 3).

Already we can loosely apply this analogy to our ontology. If we consider the world's languages to be patterns or extracted regularities from noise, then the job of the linguist is to locate redundancies in the total raw data and thereby find regularities. So far so good. Consider a sound spectograph of any sentence in any language. This visual mapping of the energy produced during speech would not specify word breaks, sentences, or phrases in that language. But those regularities are in the noise and speakers (or listeners) can discern them. Or more simply, consider speech in a foreign language. Unlike speakers of that language you won't be able to determine the meaningful boundaries and the entire utterance sounds unorganised. You are unable to compress the data. If someone asks you what was said by the foreign language speaker, your only recourse would be verbatim repetition (if you are phonetically adept enough to be able to do so). This is what Dennett seems to be suggesting when he states that a pattern "is real—if there is a description of the data that is more efficient than the bit map, whether or not anyone can concoct it" (1991: 34). The linguistic patterns are real even if they are undetectable by those unaccustomed to recognising them. Look at the following four sentences.

1. Kedi minderin üzerinde.
2. Kötturinn er á mottunni.
3. Die kat is op die mat.
4. The cat is on the mat.

Equipped with English, one can discern a pattern from (1)–(4) but each language adds slightly more distortion to the signal. (1) is in Turkish and almost indiscernible to an English speaker while (2) and (3) are in Icelandic and Afrikaans respectively, which are both Germanic languages with the latter resembling English in lexicon and syntax quite strongly. The effect is amplified in speech where the Icelandic pattern would then become quite hard to discern as well. I use these examples quite loosely here since the patterns that will concern us going forward are more structural (where language family is not always a good indicator of structural similarity).

What we need is guidance on how to compare and contrast patterns. This is where Dennett's original framework is rather unhelpful. Recall that Dennett's insight concerning patterns is that they are any aspects of a dataset that allows it to be compressed. But this doesn't give us much to go on, especially if we want to identify specific patterns and/or compare them to others. What is it about a particular set of patterns that make them pick out the same language? In addition, Dennett's original framework is deeply ambiguous on the distinction between abstract and concrete existence of patterns (see Ross 2000). Fortunately for us, there has been some development of these ideas, significantly by Ladyman and Ross and then more recently by Burnston (2017), Millhouse (2019, 2021), Suñe and Martínez (2021), designed to fine-tune this overall picture and apply it to other domains. We'll consider some of this literature below (Section 4.2.3) but in a direction less constrained by the statistical nature of information theory and more aligned to the discrete mathematical aspects of formal grammars.

4.2.2 Formal Patterns

Many, following Chomsky (1956), in the theoretical linguistic community have long argued that languages are not to be described using the tools of continuous mathematics. Grammatical theory has since drawn mostly from proof and model theory respectively. Although this picture is being eroded to some degree by the performance-based successes of computational linguistics, the competence model of Chomskyans still places a firm barrier on how far such analysis can go in principle. Before considering some useful elaborations of Dennett's real patterns framework (in the next section), I will take a brief detour into another parallel literature on pure mathematical patterns that closely mirrors the above considerations. From this exercise, I hope to lay the groundwork for a distinction between a structure or *formal* pattern and a *real* pattern, and eventually how they are related to one another for the case of linguistics.

The motivation behind modern mathematical structuralism can be traced back to Benacerraf (1973) and the dilemma he presented therein. The core idea of this foundational picture in mathematics is that mathematics is a theory of structures and systems of these structures. In this way there is a shift from the traditional (perhaps) Fregean concept that numbers, sets and other mathematical entities are abstract objects, unencumbered by spatial or temporal properties. The core insight is that it is structures and not objects which are the vehicles of mathematical truth (and knowledge). This presents an entirely different conception of the nature of the enterprise as well as the concept of a mathematical object itself. Structuralism is a broad framework with historical antecedents ranging from the Bourbaki group and Dedekind to Hilbert and even Benacerraf himself. Thus, there are a number of varieties of the idea at work within the contemporary philosophy of mathematics. I will try to stay as broad as possible for the moment, although I do plan to endorse and develop a particular variety of what is referred to as *ante rem* or noneliminative structuralism for linguistics.

In order to understand the broader features of this view on the foundations of mathematics, we need to answer a few preliminary questions. Firstly, what are structures on this view, and how do they relate to traditional objects of mathematics? Secondly, whatever they are, how do we come to know about them?

Shapiro starts his influential book with the slogan 'mathematics is the science of structure'. He continues by way of example,

> The subject matter of arithmetic is the *natural-number structure*, the pattern common to any system of objects that has a distinguished initial object and a successor relation that satisfies the induction principle. Roughly speaking, the essence of a natural number is the relations it has with other natural numbers. (1997: 5)

This holds true for groups, *topoi*, euclidean spaces and whichever mathematical structure is studied by mathematicians. Let us focus on the natural-number structure for a moment and consider its objects. What is a number on this view? Essentially, it is nothing more than a place in a natural-number structure. The only way to talk about the number 2 or 5 or 4892001 is with relation to other places in that structure, i.e., 2 is the successor of the successor of 0 or the number 2 is the third place (if we start from 0 as Frege did) of a natural-number structure, it is in the second place of an even-number structure and the first place of a prime number structure and so on. The same holds for other mathematical objects, the idea being that these objects are only interpretable in accordance with some background theory. As Parsons puts it, "the idea behind the structuralist view of mathematical objects is that such objects have no more of a 'nature' than is given by the basic relations of a structure to which they belong" (2004: 57).

The concept of a group is often taken as a canonical example of a structure. A group G consists of a finite or infinite domain of objects and a two-place function called the group operation. This function satisfies four properties (or axioms). It is associative or has the associative property, there is some identity element or the identity property, it is closed—closure property—and every element in the domain must have a reciprocal or inverse—inverse property. Now there are many different types of groups which mathematicians may wish to study. We could look at finite groups (groups with finite domains) or Abelian groups (groups whose elements are also commutative). The basic group structure is the same and the structure is given to us by the relations its objects have to one another (according to the four properties). The objects themselves are of no importance to us, they might as well be point-particles, martians, jelly-beans, or rice-crispies, it doesn't matter.[7] What matters is the structural relations one object (whatever it is) has to another in the group. We only care about the structures. In fact, we can even talk about structures in isolation from any objects. Shapiro characterises his own position in the following way:

> The first [*ante rem* structuralism] takes structures, and their places, to exist independently of whether there are any systems of objects that exemplify them. The natural-number structure, the real-number structure, the set-theoretic hierarchy, and so forth, all exist whether or not there are systems of objects structured that way. (1997: 9)[8]

So far, we have looked at the question of what structures are and what traditional mathematical objects are within them, i.e., merely places-in-structures devoid of individual meaning or significance. However, one of the characterising features of *ante rem* structuralism is that the role of the mathematical object is not discarded completely (as in the case of modal structuralism for instance, see Hellman 1989). This view allows us to retain the usual mathematical parlance involving mathematical objects (what Shapiro

(1997) calls 'realism in ontology') while still maintaining a structuralist interpretation of mathematical practice in general. The strategy employed to achieve this aim consists in appreciating the dual role objects play as positions in structures as both the more abstract 'places-are-offices' and the more common 'places-are-objects' perspective involved in quantification.

> Clearly, there is an intuitive difference between an object and a place in a structure—between an office and an officeholder. The ante rem structuralist respects this distinction but argues that it is a relative one. What is an office from one perspective is an object–and a potential officeholder–from another. In arithmetic, the natural numbers are objects, but in some other theories natural numbers are offices, occupied by other objects. (Shapiro 1997: 11)

For this reason, I think, noneliminative structuralism which contains a notion of a place in a structure as a *bona fide* object over which quantification is possible, in a model-theoretic sense, and a more abstract conception of the role (*qua* office) a position plays is a good model for retaining the object level talk present in some linguistic disciplines at the formal level.

The last aspect of this kind of structuralism with which I want to deal here is that of the relationship between actual collections of objects or *systems* as it is called by Shapiro (1997) and the structures which they exemplify. The purported relationship between systems of objects and structures is often said to be captured by the process of 'Dedekind abstraction' (in honour of another forefather of modern structuralist thinking). The idea is that a structure is an abstract form of a system which homes in on only the structural (relational) properties of the objects of the system and nothing else. Parsons (1990) suggests that we take a system such as the von Neumann ordinals as instantiating the concept of a simply infinite structure, i.e., the natural number structure (although Parsons switches the terminology used by Shapiro). As is well-known, the Zermelo numerals would do just as well for the instantiation of the natural number structure. In fact, this is a point upon which structuralism can be seen as a marked improvement on Platonist conceptions of mathematics. The two systems (von Neumann ordinals and Zermelo numerals) are not equivalent set-theoretically, yet they can still be said to exemplify the same structure. In the parlance of the previous section, if the natural numbers form a real pattern then there are essentially (at least) two non-equivalent compressions of this pattern.[9] It is important to note, that this notion of abstraction has ontological import. Unlike in the natural sciences, where abstraction is a tool used mostly for tractability or simplicity, mathematical structures (the target of mathematics as a science) *are* purely structural in nature, i.e., they only have structural properties, on this view at least.

What we have above is the characterisation of pure structures in mathematics. But such a view won't be sufficient for modelling or representing linguistic patterns. To see why this is the case, we need to have an account of the relationship or distinction between an abstract or formal pattern (here also called structure) and a real pattern found in the physical world. This is something Dennett didn't provide for us. For this task, we'll also need to depart from that of Shapiro (and Resnik) in significant ways. Consider the following remark made by Resnik specifically concerning linguistics.

> Take the case of linguistics. Let us imagine that by using the abstractive process [...] a grammarian arrives at a complex structure which he calls *English*. Now suppose that it later turns out that the English corpus fails in significant ways to instantiate this pattern, so that many of the claims which our linguist made concerning his structure will be falsified. Derisively, linguists rename the structure *Tenglish*. Nonetheless, much of our linguist's knowledge about *Tenglish qua* pattern stands; for he has managed to describe *some* pattern and to discuss some of its properties. (1982: 101)

In linguistics we seem to be concerned with a specific class of structures, those which are instantiated in the real world. These are the structures that are produced by human linguistic competence, i.e. the outputs of competence if we follow Devitt (2006). In the mathematical structuralism literature, Parsons (1990) comes closest to describing this with what he calls 'quasi-concrete' objects. He offers the existence of such objects as an objection to pure structuralism. He states that there are "certain abstract objects that I call quasi-concrete, because they are directly 'represented' or 'instantiated' in the concrete" and he includes as an example of such an object "symbols whose tokens are physical utterances or inscriptions" (1990: 304). Delving into these debates, although fascinating, will take us too far afield from our current path. The basic problem is that pure structures are not good candidates for real patternhood. They certainly admit for the possibility of compression (I can characterise the entire infinite set of natural numbers with a few axioms and a successor function) but fail when it comes to the application of non-redundancy tests (more about this in the next section). *Pace* Platonists, languages are not pure mathematical objects (or structures).

What we've seen so far is an intuitively appealing (albeit information-theoretic) account of real patterns in the previous section. But it didn't provide us with much traction on how to access the similarities and discontinuities between various compressions of the patterns we encountered. We then moved on to a different philosophical terrain where patterns *qua* structures have featured prominently. This framework compensated for the previous one by bringing all the rich resources of mathematics and set theory to bear on issues of structural similarity and characterisation. But it failed to give us clear guidance on how to characterise patterns which are found in the world. The next section can be seen as an attempt to marry the two concepts and essentially provide an answer to the question: *How are formal structures realised in real patterns?*

4.2.3 Real Patterns Revisited

One might be tempted at first glance to resurrect the old type-token distinction for the present task. Formal patterns are types and real patterns are tokens of these types. But as I have mentioned earlier, the intuitive appeal of this distinction is quickly lost upon further scrutiny. An issue with which Dennett himself wrestled was that the physical reality of real patterns is not obvious. Some are like physical tokens and apparent to even the human eye (his barcode example) but others are indiscernible yet presumably still there. In addition, Benacerraf-type worries threaten any neat characterisation between formal types and their physical tokens, if that former share no spatio-temporal features with the latter (or have none to begin with). In this section, I'm going to motivate some tools for discretely characterising real patterns before modifying Ladyman and Ross's 'scale relativity of ontology' thesis as a replacement for the ever recalcitrant type-token tendencies.

The key to Ladyman and Ross's modified theory of real patterns is the notion of 'indispensability'. Recall that Dennett wasn't very clear on which aspects of datasets enabled compression and thus pattern recognition. Ladyman and Ross proffer some direction in response. They claim that these aspects are patterns which partially encode the original dataset. This generates multiple 'mutually redundant' patterns in that many of the patterns might map on to or represent the exact same aspects of the dataset. For them, a pattern *simpliciter* is "just any relation among data" (Ladyman and Ross 2007: 228). A theory of real patterns should help us with choosing which patterns (partial encodings) are the real ones from sets of mutually redundant options. So the idea is that the indispensable patterns are the real ones. It's important to note that redundant patterns can provide useful information, according to them, but can be ultimately discarded without ontological loss in the description of the dataset. Their final definition looks as follows:

> To be is to be a real pattern; and a pattern $x \rightarrow y$ is real iff
> (i) it is projectible; and
> (ii) it has a model that carries information about at least one pattern P in an encoding that has logical depth less than the bit-map encoding of P, and where P is not projectible by a physically possible device computing information about another real pattern of lower logical depth than $x \rightarrow y$. (Ladyman and Ross 2007: 233)

(i) above involves *projectibility*, which is a relation that ensures that everything that is needed to compute y is present in x or in algorithmic complexity terminology 'y is projectible from x' means that there is "a trivial (very short, etc.) universal Turing machine running x as its program outputs y" (Suñe and Martínez 2021: 4322). (ii) guarantees that the pattern identified encodes indispensable ontological information. It's basically an information-theoretic version of Occam's razor.

To see dispensibility in action, consider a case from the history of science. Ether was initially put forward as a substance that acted as the medium for the transmission of light through space. Einstein famously showed that ontological commitment to the ether was not necessary in the explanation of the phenomenon. Applied to the current framework, Lorentz's theory did not identify a real pattern as Einstein's did since its chief ontological postulate was dispensable in the explanation of the data.

> Lorentz's ether theory had many predictions and observations, including length contraction, which is the phenomenon that a moving object's measured length will be shorter than its proper length. Einstein proposed an alternative theory, special relativity, that has the exact same predictions and observations (Bradley, 2020, p. 9), including length contraction, without needing to posit the existence of ether. Special relativity thus showed physicists that ether is dispensable. (LeBrun 2021: 2)

The framework thus aims to exclude all patterns that do not result in a loss of information from real patternhood. Of course, we can also see that this procedure is not designed to distinguish formal or mathematical patterns, discussed in the structuralism literature above. The problem is that mathematical truths hold universally and we can't apply non-redundancy tests to purely mathematical structures. All mathematical patterns are equally real in this sense (or none are!).

More significantly, Suñe and Martínez (2021) worry that Ladyman and Ross's account too fails at its own test. The details will not detain us here but they basically show that for any pattern that carries information and is smaller than the shortest program (i.e., the Kolmogorov complexity), it is a real pattern according to Ladyman and Ross. However, this is an untoward result since "it is trivial to multiply the redundant patterns that pass the RP test by taking a real pattern and adding to it any amount of noise such that the total length is less than the Kolmogorov complexity of the original dataset" (Suñe and Martínez, 2021: 4326). Of course, this is a technicality (and they do proffer a fix in terms of conditional Kolmogorov complexity). I think there's an easier path for our purposes.

Millhouse (2021a) expands on Ladyman and Ross' proposal by extending their discussion of scientific models. For the latter, our access to patterns might require mediation by models. They are, however, careful to distinguish this situation from an identification between model selection and pattern individuation. What Millhouse adds is a similarity criteria of model fidelity (Weisberg 2016) into the standard repertoire of information theory. He claims that this allows for a more realist grounding of real patterns. From our perspective, it allows for the inclusion of mediation by models and provides a tool, namely similarity, to assess the patterns picked out by our formal models.[10] The connection between similarity and compressibility is apparent in information theory. For example, one measure of similarity (the 'Levenshtein distance') compares the number of single character edits needed to transform one series A of strings into another B. If one generalises this procedure by means of more efficient set of instructions involving multiple steps at once, we can see that similarity (according to this measure at least) and compressibility are related. "The key point is that a program for recovering A from B and *vice versa* must contain just enough bits to reproduce those elements of A that B lacks (and *vice versa*)" (Millhouse, 2021a: 6). In other words, the 'shortest program' that maps A to B (and the other way around) would encode the difference between them similarly to our definition of compressibility above. Therefore, measuring or considering the similarity between formal models of particular patterns can yield information about compressibility and thus real patternhood.

Furthermore, models can be faithful to real patterns in at least two ways. The first involves how the output of the model tracks the output of the real world in terms of predictions made. Weisberg (2016) calls this 'dynamical fidelity criteria'. We'll consider this topic in Chapter 7. The other way models represent reality is more relevant to the present discussion, namely 'representational fidelity criteria' or "whether the structure of the model maps well onto the target system of interest" (Weisberg 2016: 267). One issue (pointed out by Millhouse) with Dennett's analysis of real patterns is that it is unclear that it can explain how the structure proposed by a given model can be realised by physical systems which are dissimilar in patterning. We'll address a specific version of this type of mulitiple realisibility in the next section.

On the flip-side, models come in different shapes and sizes. Essentially they allow for different angles on the target phenomenon to be emphasised and homed in on. Millhouse describes Dennett's position as:

> Essentially, a model of a physical system (like a human brain) can be cast at different levels of abstraction. To describe the operation of the brain (and hence to account for its behaviour) at the microphysical level would require a vastly complicated model. Fortunately, we can also adopt a simpler, high-level model. (2021a: 3)

This idea dovetails with Chomsky's repeated claims about I-languages existing at a different level of abstraction (see Chomsky 2000a). Again, we will return to a more formal framework for capturing this insight in Chapter 7. But for now we can think of it this way: linguistic models operate at the simpler, high-level of description involving mental states. The patterns they identify are no less real for not being associated directly with neuronal states (or quantum particles for that matter). This brings us to the final element of the account, the thesis that will offer us an alternative to type-token distinction for linguistic ontology. This is what Ladyman and Ross call the 'scale relativity of ontology'.

To begin with, we need to appreciate the fact that the type-token distinction is part and parcel of a larger metaphysical and anti-naturalist framework for reality, i.e., the "commitment to a world of levels strictly composed out of deep-down little things" (Ladyman and Ross 2007: 193), i.e., that types are composed of or instantiated by their mundane tokens. The most common expression of this framework can be found in the many familiar supervenience theses scattered throughout the philosophy of mind, aesthetics, and general metaphysics. Ladyman and Ross single out the work of Jaegwon Kim but he is by no means alone in this. The alternative idea is something of the following sort:

[I]t is the idea that which terms of description and principles of individuation we use to track the world vary with the scale at which the world is measured [...] Here are some more: at the quantum scale there are no cats; at scales appropriate for astrophysics there are no mountains; and there are no cross-elasticities of demand in a two-person economy. (Ladyman and Ross 2007: 199)

Instead of using supervenience to account for emergent properties and the relationship between different ontological categories or explanations, they introduce a radical idea that ontology is scale relative. In other words, within different levels of theory, as well as different length and time scales, distinct structures of causation and law emerge. It's a bit like looking through multi-layer glasses with each layer picking up different forms of light many of which, such as x-rays, ultraviolet light, infrared radiation, and radio waves, are invisible to the human eye. By removing successive layers, we can witness different realities all equally existent. Thus, the special sciences identify some structures that are not immediately apparent in fundamental physics.

As we've seen above, Dennett (1991) gets off the instrumentalist bench and describes an ontologically motivated view. A *real pattern* (RP) is something that provides a simplified description relative to some background ontology. Specifically, "[a]ccording to RP, the utility of the intentional stance is a special case of the utility of scale-relative perspectives in general in science, and expresses a fact about the way in which reality is organized—that is to say, a metaphysical fact" (Ladyman and Ross 2007: 199). For example, to a psychologist schizoaffective disorders are identified and predicted in terms of clusters of sets of behavioural dispositions, to a neuroscientist brain structures, chemical imbalances and genetic predispositions feature more prominently. Perhaps psychiatrists admit both intentional states and chemical processes into their ontologies. Good examples abound.

Real patterns exist at the material level and structures at the formal.

'Real patterns' should be understood in the material mode. Then 'structures' are to be understood as mathematical models – sometimes constructed by axiomatized theories, sometimes represented in set theory – that elicit thinking in the formal mode. (Ladyman and Ross 2007: 119)

The appreciation of this distinction is why we need to banish the type-token distinction. The relationship between these modes is not some one-to-many instantiation relation, rather it consists in a "claim that successful scientific practice warrants networks of mappings as identified above between the formal and the material" (Ladyman and Ross 2007: 121). Formal structures *represent* real patterns. These could be the *indirect* surrogative kind of representation characteristic of scientific modelling as Ladyman and Ross state. In some cases, this relationship can be captured by measurement (see example below). Towards the end of this chapter, we will relate grammars to models in precisely this way. In this case, failure to pick out a real pattern often involves identifying only purely formal patterns (where, in our case, a grammar fails to find an actual linguistic pattern to represent).

Transposed to the case of linguistics, at one level—that of mathematical modelling—we see patterns of regularities which resemble formal rules, and at another we observe the behaviour of individuals producing and grasping natural language structures—the psychological level—and at yet another level we see linguistic regularities emerging within populations of individuals, the sociological level. For example, many linguists consider language to be infinite or constitute a discretely infinite set, in the formal and generative literature. But of course to sociolinguists, languages are unequivocally finite in that every actual utterance, past and present, of every human language clearly forms a finite collection of objects. In the next chapter, we will see a similar ontological shift from individual organisms to biological systems of multiple organisms coupled with their environments within systems biology.

Consider the following explication of the core idea. Natural language (NL) is a non-redundant RP because (among many other things) it describes an uncountable range (or discretely infinite set) of future languages that the conjunction of models of existing languages does not. Let us consider this concept in terms of measurement or some features related to predicting grammaticality (according to some formal linguistic theory). Empirically (and as incorporated in linguistic theory), every measurement of isiXhosa (or Swiss-German) is a measurement of NL, but not *vice-versa*. isiXhosa is then also a non-redundant RP because it is has some special, predictively crucial, measurable features. If it turned out that isiXhosa and isiZulu had been statistically or formally indistinguishable from one another with respect to the variables and parameters of linguistic theory, then we would have discovered that they are in fact the same language, at least syntactically. Each would have turned out to be a redundant RP, and we would conclude that we had a single RP with two different names used by different groups of people.[11] But of course, as we will see in Chapter 8, there are a number of linguistic levels at which measurements can be taken and thus a number of divergent patterns onto which we can latch for language identification. For instance, Mandarin and Cantonese are written in very similar ways but are not mutually intelligible from a phonological perspective. In a different vein, Ukrainian and Russian are very similar grammatically but differ in lexicon.

The central idea of the present section is that there are such things as real patterns in nature, and natural languages can be considered to fit this description at different levels. There are no abstract languages somehow instantiated by their physical tokens. However, the task of providing some account of the relationship between linguistic patterns and our formal tools for identifying them remains. In other words, how do grammars identify *real* linguistic patterns?

4.3 Grammars as Compression Algorithms

In this section, I want to proffer a particular interpretation of what grammars are in line with the real pattern analysis of the chapter so far. In the literature, grammars have been described as scientific theories of language (Chomsky 1986), functions (Lewis 1969), and formal models (Tiede and Stout 2010). All of these characterisations aim to capture some element of the role grammars are meant to play in the theory of language and how they map onto linguistic reality. In other words, the challenge is to show how grammars function and what their ontological import is. My answer to this challenge is simple: *grammars are compression algorithms*. There are two salient elements of this characterisation, one relates to how they pick out patterns in linguistic data and the other relates to how they identify linguistic structures in the world.

There is a common idealisation at the core of both the application of complexity theory to real patterns and the advent of generative grammars, one that is recognised by Ladyman and Ross for the latter. As Suñe and Martínez (2021: 4322) state: "[i]n order to apply algorithmic complexity theory to real-world phenomena, we need to translate [patterns] into strings". In formal language theory (FLT hereafter) the branch of mathematics invented to study natural language structure, the initial idealisation similarly idealises sentences as sets of strings.

In the following section, I will start by showing how grammars can be viewed as compression algorithms. But this will only get us to the data level of patterns, i.e., patterns in the data. To ascend to the level of linguistic ontology, we need to consider the structures that come with the grammars. An important caveat:

the forthcoming only concerns syntactic patternings at this stage. We will add to the tools of measurement and real pattern detection in Chapter 8.

What is a compression algorithm? The basic idea is that a compression algorithm is a procedure for creating compressed representations of the data from which the original data set can be recovered. As Millhouse (2021b: 25) puts it: "when we say that C is a compressed representation of D, we mean that C is considerably smaller than D and that there is some practical method for generating C given D and *vice versa*". This is essentially what a grammar does for a language. It is a compressed representation of the linguistic data. A grammar of English is meant to generate the sentences of English, usually *via* a finite set of rules.

Indeed, compressibility was always central to the formal language theory enterprise. Consider a recent description of the advent of the field and its concept of a formal grammar:

> [T]his new branch of mathematics provided the formal grounding for a new conception of linguistics in which *grammars*, rather than sentences or collections of sentences, were the scientifically central objects: instead being derived from collections of sentences as compact summaries of observed regularities, grammars are seen as (ultimately mental) systems that determine the status of sentences. The "observed regularities" come to be seen as consequences of the structure of the underlying system, the grammar. (Hunter 2021: 74)

In other words, FLT is in the business of defining structural depictions (*via* grammars) that generate the observed regularities of datasets as a consequence, instead of just collecting large datasets. One means of capturing this function of a grammar vis-à-vis linguistic regularities is by appreciating its role as a *set enumerator*. If sentences can be modelled as strings and languages as sets of these strings, then a generative grammar specifies the finite rule system that produces or *generates* that set. The game is more specific than this. What we want is the most compact means of enumerating the members of the set. The most broad notion of set enumeration is equivalent to a Turing machine in which the set of recursively enumerable languages is produced by a special kind of automaton or enumerator. But this notion is too broad since it maps onto a level of formal complexity which characterises any computation. We are interested in a much smaller subset of real patterns.

A better way of thinking about the kinds of compression algorithms we are dealing with is the idea (courtesy of Geoff Pullum) that a generative grammar is a *nondeterministic random enumerator*. This formal interpretation is meant to demarcate the real patterns from the noise. We'll get to this in a moment.

Before we can continue with the discussion at large, we need to appreciate some of the preliminary idealisations involved in FLT. Formal languages are composed of strings which are represented by means of sequences of letters: $\{a, b, c, d, \ldots, y, z\}$. This is called the *alphabet* over which the language is defined or Σ. Letters are used for convenience of presentation but they could be any symbol including numerals (or binary code).

There are a few useful operations on strings, such as concatenation, exponentiation and reversal that can already model basic derivational morphology. For present purposes, the notion of a substring is important. "Given a string ω, a *substring* of ω is a sequence formed by taking contiguous symbols of ω in the order in which they occur in ω" (Wintner, 2010: 12). Then a substring is defined iff there exists strings ω_1 and ω_r such that $\omega = \omega_1.\omega_c \cdot \omega_r$ (where '·' is string concatenation). Two particular kinds of substrings are especially useful, *prefix* (ω_1 above) and *suffix* (ω_r above) respectively. These are essentially compressions of the superstring. For example, if ω=*compositionality* then the empty string (denoted by ϵ), *co, comp, composit,* and *compositionality* are all prefixes. The suffixes are ϵ, *y, ity, tionality,* and *compositionality*.

Finite state grammars (and the automata with which they are associated) do a good job of identifying patterns by means of prefix compressions or what Hunter (2021) calls the 'forward set'. "The forward set of a string s is the set of states that the automaton could reach from its initial state by taking some sequence of transitions that produce s" (Hunter, 2021: 80). Basically, the spot in the machine that accepts the sequence generated so far. If the transitions are words a,b,c, and the states are non-terminals (or syntactic categories) like NP, PP, VP, etc. then the forward set of some state is the one that realises the sequence. In the concrete example below, the forward states of *Irene is* are both S2 and S3 and similarly the forward state of *Irene is very* is only accepted by S3.

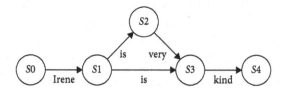

Armed with the forward set of some prefix, we can calculate the forward set of the entire string. In other words, a substring gives us enough information for identifying the pattern of the entire sequence. In this sense, the string is essentially compressed by the forward set. Not only this but two substrings with the same forward set are substitutable for one another which also allows us to identify the shortest path or forward set which produces the entire string. This would get us some way to spotting the real patterns in the stringset.

It is a bit premature to discuss real patterns since we haven't defined a language explicitly yet. Languages are sets of strings or stringsets over some alphabet Σ (to be precise, it is a subset of the infinite Kleene closure of Σ or Σ* which is an infinite set). There are a number of mechanisms for defining formal languages. Such mechanisms, or linguistic formalisms, are composed of rules, and a rule is a non-empty sequence of symbols or a mixture of terminals and non-terminal symbols (in CFGs below). Sets of finite rules for defining stringsets are also called *grammars*. Putting all of this together we get the following, where S is the start symbol:

> G will be said to *generate* a string w consisting of symbols from Σ if and only if it is possible to start with S and produce w through some finite sequence of rule applications. The sequence of modified strings that proceeds from S to w is called a *derivation* of w. The set of all strings that G can generate is called the *language* of G, and is notated $\mathcal{L}(G)$ (Jäger and Rogers 2012: 1957).

Furthermore, and more to the point, grammars identify the patterns in the stringsets which model natural language syntax. Once these patterns are identified, then we have some traction on the language $\mathcal{L}(G)$ of which they comprise. "The task of designing a grammar to generate some desired pattern amounts to choosing which distinctions to ignore and which distinctions to track" (Hunter, 2021: 75). In other words, grammars identify the structures which map onto real patterns and those which only involve noise. This is the essence of a random set enumerator discussed above. Or as Sells puts it:

> A generative grammar consists of (i) a set of initially given strings available at the start, plus (ii) a set of operations for constructing new strings from strings that are antecedently available. A generative grammar defines a given language if and only if every sentence of the language, and only those, can be constructed from the initially given strings using the operations. (2021: 246)

Another way of characterising this concept is that the grammar generates some language L (i.e., $\mathcal{L}(G)$) iff for every well-formed string x in L there is some way of using the grammar (set of rules) to construct a derivation ending in x and there is no such procedure for ill-formed sequences. This shows that the task of the grammar is to formally model the patterns of the language. In an almost Dennettian claim, Hunter brings out the relation between compressibility and grammars in the following statement:

> [D]esigning a grammar amounts to choosing "what to remember" about intermediate subexpressions, and what can be ignored. It is exactly this move of ignoring distinctions between subexpressions that allows a finitely specified grammar to generate unboundedly many expressions. A device that never allowed itself to "forget," or ignore, some aspects of an expression's internals, would be one where the applicability of a grammatical rule is dependent on the *entire* surrounding context, and would simply amount to a finite list of expressions [...] This set exhibits no interesting patterns because the automaton has no interesting structure. (2021: 83)

Such a set would be random and akin to the entire incompressible bitmap of information discussed earlier. My claim is that grammars *qua* random set enumerators, equipped with various rules, which are essentially compression algorithms, search for real patterns within the language (which itself is a real pattern).[12] This is the crux of the compressibility of grammars.

But so far, we only have traction on how grammars might identify or compress patterns *simpliciter*. We have no direct way to get from the patterns to the structures in reality. Thus, what we need is an analogue of Ladyman and Ross's indispensability criterion to get us closer to real linguistic patternhood.

The property of indispensability is related to the issue of multiple realisability mentioned in the previous section. If grammars are formal models of stringsets then how do we distinguish between distinct grammars which generate the same stringsets? Or more specifically, how do we identify the real patterns from the models, if the models pick out the same sets of sentences?

In order to see how indispensability prefigures into this scenario, we need to appreciate the difference between *weak* and *strong generative capacity* (Chomsky 1963). Two grammar formalisms are said to be weakly equivalent if they model the same sets of strings. In other words, if for a grammar of the first type of formalism there exists a grammar of the second type that characterises the same sets of strings, then the grammars are said to be weakly equivalent (or possess the same weak generative capacity).[13] There have been a number of proofs showing that many grammar formalisms previously thought to be distinct are in fact weakly equivalent (see Vijay-Shanker and Weir 1994). Therefore, the possibility of multiple grammar formalisms which enumerate the same sets poses a *prima facie* problem for Ladyman and Ross's criterion of indispensability of real patterns. Put in slightly different terms, the grammars might be mere 'notational variants' of one another (Chomsky 1972) or redundant compressions of the dataset (again see Chapter 5 for some discussion of this concept). In other words, if all the models are picking out mutually dispensable patterns then we won't get any traction on the real ones that characterise natural language structure.

To see how to get from the data to the structures we need to define indispensability for grammar formalisms. For this, we need to consider strong generative capacity. Chomsky himself suggests a stronger notion is needed to capture linguistically important *explananda* like acquisition data.

> [D]iscussion of weak generative capacity marks only a very early and primitive stage of the study of generative grammar. Questions of real linguistic interest arise only when strong generative capacity (descriptive adequacy) and, more important, explanatory adequacy become the focus of discussion (1965: 61).

Informally put, each grammar formalism imposes a certain structure on the sets of strings of the languages they characterise, sometimes called a 'structural description'. For instance, phrase structure grammars, most commonly used in generative linguistics and a special case of Context-Free Grammars (one level up from the finite-state grammars we've been talking about), assign hierarchical tree structures to the strings. If two grammars are strongly equivalent, then they assign the same structural descriptions. To be strongly equivalent, weak equivalence needs to be coupled with structural equivalence or equivalence of structural description. It turns out that the proofs concerning the equivalence of grammar formalisms such as phrase-structure, dependency grammar, categorial grammar, and tree-substitution grammar are *only* located at the level of weak generative capacity since each formalism assigns different structural descriptions to the strings. Compare the two structural descriptions

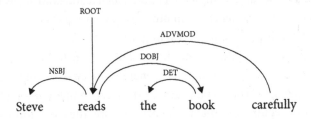

Fig. 4.1 Dependency graph for *Steve reads the book carefully*.

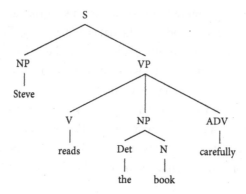

Fig. 4.2 Phrase structure tree for *Steve reads the book carefully*.

generated for the string *Steve reads the book carefully* by a dependency grammar and a phrase-structure respectively.

Dependency grammars produce rooted directed acyclic graphs while phrase structure grammars produce (usually) binary trees (a kind of rooted connected graph). Each model carries indispensable information about the structures underlying the stringsets despite generating the same strings. Dependency grammars are often said to represent argument structure more perspicuously while phrase structure captures hierarchical constituent relations to greater effect. Thus, each grammar formalism provides vital and indispensable onto-logical information about the language *qua* real pattern. Where this is not the case, the less informative grammar is redundant. A case of redundancy is given by Tree-substitution grammars (TSGs) and Tree-adjoining grammars (TAGs). Both are weakly equivalent. However, TAGs also capture all the structural information provided by TSGs but not *vice versa* (Rambow and Joshi 1997).

With this in mind, we can characterise indispensability as the features of a grammar formalism or model that are *not* strongly equivalent to features of other models of the same stringset or pattern. Such features are likely to identify the structure of the real patterns. In other words, strong generative capacity is the key to the indispensability of real linguistic patterns.

On a somewhat technical note, finite state automata like the one above (and their corresponding grammars) generate the class of languages known as regular languages. These languages are the least complex among the classes (ignoring sub-regular for the moment) and are thus the most efficiently parsable. They are also closed under a number of useful operations (like union, intersection, exponentiation, complementation, etc.). Although there are many interesting applications of these grammars to natural language processing, we will not be directly interested in those here. For one thing, it is a well-established fact that regular languages and finite state grammars are inadequate as characterisations of natural language syntax (Chomsky 1956, 1957). Basically, certain kinds of patterns found in natural language require rules that regular grammars cannot produce. One such pattern is $a^n b^n$ where the n number of a's are followed the same number of b's. This is because (informally) the grammar would have to remember this number n while waiting for the end of the b's. Many natural language patterns involve such patterns. *Either ... or* and *if ... then* constructions are prominent examples (see Miller and Chomsky 1963). In order to capture this kind of potentially unbounded 'nesting', we need Context-Free Grammars (CFGs) with recursive rules such as $S \rightarrow aSb$.

The key to appreciating how finite state languages fail to capture these linguistically relevant patterns and CFGs improve upon them, brings us back to our initial forward sets defined in terms of prefixes. Prefixes can't compress string sequences or patterns which require the infinite collection of prefixes to be separated by some function "that maps each prefix to a subset of a finite number of states" (Hunter 2021: 84). This is the role performed by the nonterminal S symbol of CFG rules. From a compression point of view, *infixes* are needed over and above prefixes. Whereas prefixes are used to determine the pattern that comes after, infixes classify strings in terms of the material around them. This expands our recognition tools to *inside sets* which essentially split strings into substrings in abitrary ways. A CFG rule is a combination of terminals and non-terminals such that the first element is always non-terminal and the second some sequence of symbols. There are more interesting features of these rules involving endocentricity and so on that will be addressed in Chapter 7.

For our purposes, we can see that the progression from formal patterns or structures captured by finite state grammars only partially pick out real linguistic patterns. To do better we need to find ways of characterising common and more structural properties of strings. In fact, in their seminal work, Hopcroft and Ullman (1979) identify the central mathematical proof method of formal language theory as what they call 'structural induction'. Although this proof strategy is similar to mathematical induction it operates over the domain of recursively defined structures as opposed to that of the positive integers. Its validity as a proof strategy is based on the validity of mathematical induction *simpliciter* where "[l]ike inductions, all recursive definitions have a basis case, where one or more elementary structures are defined, and an inductive step, where more complex structures are defined in terms of previously defined structures." (Hopcroft et al. 2001: 23). You can think of these elementary structures as trees. So the induction runs by firstly proving that some property P holds for the base case of the tree (this could be the empty tree or a simple subtree or the root node), then *via* the inductive hypothesis assumes that $P(T)$ for some arbitrary tree. Finally by the induction step, we show that $P(T')$ for some new tree T' derived from the subtrees T_i. Again structure, not just compression, is paramount to the enterprise.

Importantly, I'm not saying that CFGs capture the structure of natural language syntax. In fact, the consensus is that they do not, based in part on proofs from cross-serial dependencies found in Swiss-German and Dutch (Shieber 1985, Bresnan et al., 1982). Whether some more complex or 'mildly context-sensitive' grammar formalism performs better is a matter of debate (Joshi 1985).[14] Nevertheless, our purpose here was to establish two points about formal grammars, firstly that that can be considered compressions of linguistic patterns and secondly that identifying the real or indispensable patterns among the many formal ones involves serious structural considerations.[15]

This leaves us with the following definition of a 'real pattern' (modelled on Ladyman and Ross' general analysis):

> To linguistically *be* is to be a Linguistically Real Pattern (LRP); and a linguistic pattern $x \rightarrow y$ is real iff
> (i) it is grammatically projectible or there is a nondeterministic random enumerator which generates y from x; and
> (ii) it has a model that strongly generates at least one language L or string ω where L or ω is not grammatically projectible by a physically possible grammar computing information about another LRP (L') of lower logical depth than $x \rightarrow y$;
> (iii) it has at least one unique graph-theoretic structure which corresponds to the stringset generated[16]

Although this might look complex, the basic idea is not meant to be. In order for something to be a linguistically real pattern, i.e., a language, it needs to involve only the sentences of that language and its grammar needs to be a compressed representation of that set as well as identify indispensable structures associated with the patterns of the set. Finding these patterns and structures can involve various forms of measurement such as predicting future grammatical strings or showing why certain strings are ungrammatical (more prevalent in model-theoretic syntax). The last two conditions, (ii) and (iii), are just two ways in which we can specify the structures involved beyond the compressions of the data.

With the formal framework for demarcating the linguistically real patterns from redundant patterns and noise in hand, let us move on to the ontological question of where we should begin to look for these patterns in nature.

5

Linguistic Patterns and Biological Systems

The next part to the puzzle involves answering the question 'real patterns of what?' In other words, what are these patterns, and the grammars that characterise or measure them, really tracking? The short answer is *structures*. The long answer is *the structures of specific kinds of biological systems*. At this junction, some of the other ontological options resurface (as well as new ones).[1] They could be tracking mental states, idealised linguistic tokens, or linguistic conventions. In the following chapter, I hope to show that all of these answers have some plausible contribution to make but do not suffice individually. Specifically, I offer a novel approach to understanding linguistic ontology by reconceiving the idealisation of a 'linguistic community' in terms of an analogy with a biological system. Thus, the central insight of the present chapter can be summarised as follows:

CENTRAL INSIGHT IV: NATURAL LANGUAGES ARE EMERGENT PHENOMENA WITHIN DYNAMIC COMPLEX BIOLOGICAL SYSTEMS COMPRISING NETWORKS OF INTERNAL MECHANISMS, EXTERNAL CONVENTIONS, AND ENVIRONMENTAL FACTORS.

After making my case, I will briefly suggest its possible extension to issues of language acquisition in terms of network analysis. I should note that the inclusion of the Chomsky quote from the epigraph was slightly cheeky. In fact, in that passage, Chomsky goes on from describing language as a biological system with all of its complexity and messiness to advocating a view of language as optimally designed and minimalistically simple. Or at least, competence is clean and performance is messy. He insists, there and elsewhere, that language is unique or an outlier in the biological world for its near-perfection. But why the controversial gambit?[2] Why not embrace the complexity of biological systems? One reason is that complexity science is an incipient field of inquiry with massive disagreement on what complexity is and which systems exhibit it (see Ladyman and Wiesner 2020).[3] But new insights are constantly emerging and certain perspectives on biological systems are reasonably well-established. Hence, I plan to harness these insights into an analysis of 'biolinguistics' in

Language, Science, and Structure. Ryan M. Nefdt, Oxford University Press.
© Oxford University Press 2023.
DOI: 10.1093/oso/9780197653098.003.0005

terms of systems biology, or what I call *systems biolinguistics*, and thereby ground the real pattern analysis of the previous chapter within a newly succoured concept of the *linguistic community*.

5.1 Biolinguistics and Biology

If you will recall from Chapter 2, Chomsky made a strong case for the abandonment of any hope of a respectable scientific object of study based on externalised or E-languages. Instead, language is to be conceived of as an *individualised* internal computational state. At various epochs of the generative tradition in linguistics the field has aligned itself with psychology. More recently, however, the claim of biological correspondence has been more pronounced especially in the so-called biolinguistic programme (Lenneberg 1967; Chomsky 1995; Boeckx 2006).

> The biolinguistic perspective views a person's language as a state of some component of the mind, understanding 'mind' in the sense of eighteenth century scientists who recognized that after Newton's demolition of the only coherent concept of body, we can only regard aspects of the world 'termed mental' as the result of "such an organical structure as that of the brain". (Chomsky 2005: 2)

Notice despite the change in nomenclature the claim remains that language is a state of the mind - the so-called language faculty. Linguists in the generative tradition have often made claims about natural language as an organ of some sort due to their adherence to the modularity of mind or following the innateness hypothesis, metaphors have been used to suggest that language 'grows in the mind'. Anderson and Lightfoot (2002) take this picture very seriously:

> We conclude that there is a biological entity, a finite mental organ, which develops in children along one of a number of paths. The range of possible paths of language growth is determined in advance of any childhood experience. The language organ that emerges, the grammar, is represented in the brain and plays a central role in the person's use of language. (22)[4]

And of course, since the *Minimalist Program* (Chomsky 1995), evolutionary concerns have been taken as constitutive of the theory with both

Hauser et al. (2002) and Berwick and Chomsky (2016) arguing for a saltation based origin story for language. Chomsky has claimed that "[w]e study these objects [languages] more or less as we study the system of motor organization or visual perception, or the immune or digestive system" (1999: 33).[5] To my knowledge, the digestive system is not studied in terms of formal grammars (although the immune system has, see Jerne (1985)). In fact, it is questionable how far this biological analogy goes beyond the metaphorical. As Boeckx and Martins state:

> Biolinguistics has in practice been seen as a sub-field or rebranding of generative linguistics, and as such most of the work said to be biolinguistic came from there. But the biology is rarely taken seriously into account in such work: even though linguistics managed to incorporate biology into its rhetoric, it has remained a largely descriptive field, with rare signs of real biological work. (2016: 4)

Even if we do take the talk of biological organs and the like seriously, linguistics under the biolinguistics banner has not really attempted to incorporate data or theories from the biological sciences within its remit. Additionally, I am particularly worried that this kind of parlance involving organs and isolated biological objects invites object-oriented ontologies which are better left abandoned. Nevertheless, I think there is a better way to understand language as biological, not as an organ but rather a complex system in terms of systems biology.

The philosopher who comes closest to this approach is Ruth Millikan (1984, 2005). In her defence of the concept of public language, Millikan challenges the heart of Chomskyan I-language. She accepts the existence of an internal state responsible for linguistic forms. She even takes onboard the idea of a language faculty. But where she departs from Chomsky is in his rejection of communication as the primary function of language. Or as she puts it "that a primary function of the human language faculty is to support linguistic conventions, and that these have an essentially communicative function" (Millikan 2005: 25). In a sense, however, she deviates from both I-language claims and E-language claims to deliver a *sui generis* perspective. She agrees with Chomsky that the boundaries between what people usually refer to as public languages such as English, German and Dutch are vaguely defined. People have strong intuitions about the distinction between Dutch and German but in a border town between the two countries, the language spoken there is allegedly mutually intelligible. Similarly, Nguni languages of

Southern Africa such as isiXhosa and isiZulu share so much syntactic and lexical structure that they too are mutually intelligible in many contexts. A common quote attributed to the sociolinguist Max Weinreich is that 'a language is a dialect with an army and a navy'. Xhosa and Zulu people tend to share the same army and navy but the point of course is that what demarcates languages in this external sense is not something susceptible to precise or scientific characterisation. Later, I will argue that external languages form a continuum (and it's not a scientific disaster despite claims to the contrary). Importantly, Millikan joins Chomsky in rejecting the idea of some 'fixed entity' which speakers of such languages are said to learn, acquire or know. Here I concur, as is evident from the dialectic of Chapter 2.

However, as previously mentioned, Millikan does not then embrace Chomsky's I-language concept by default. Focusing on the acquisition claim, she states the following:

> Learning language is not merely acquiring an 'I-language'. It is not just achieving a relatively steady state of the language faculty. Learning language is essentially coming to know various *public* conventions and, with trivial exceptions, these conventions are around to learn only because they have functions. (Millikan 2005: 26)

The key to appreciating this alternative is through disabusing the concept of 'conventions' from its usual Lewisian connotations involving regularities (*de facto* or *de jure*). For Millikan, the language faculty's primary job is to create and sustain language conventions and the job of those conventions in turn is to make communication possible. She defines conventions as activities or patterns of activities. These patterns can be exhibited by one or multiple participants. Dancing conventions form patterns that involve the interactions between partners which particular roles to play. "To become conventional, an activity or pattern of activity must, first, be reproduced, hence proliferated" (Millikan 2005: 30). The kind of conventions language produces are 'coordinating conventions'. The reproduction is partly arbitrary and based largely on tradition and not efficacy. This can explain why we see such diverse linguistic forms within and across languages. So where does the biology come into the picture? Here she relies on the concept of 'proper function' which is akin to 'survival value'.

> Roughly, the function or functions of a conventional pattern are those effects of it that account for its continued reproduction. More accurately, the pattern

is proliferated due in part to a correlation between it and certain of its effects. It is selected for reproduction, in accordance with conscious or unconscious intent, owing to its being coincident with these effects enough of the time. (Millikan 2005: 50)

Languages, on this view, are webs or networks of conventions that form concrete sets of speaker–hearer interactions "forming lineages roughly in the biological sense" (Millikan 2005: 28). These lineages are not constrained by the properties of I-languages according to Millikan. In response to this account, Chomsky (2003) denies that it gets much further than standard E-language accounts and in so far as it does, he claims it basically amounts to his I-language conception. He argues that conventions are still left vague in Millikan's theory and that the primary function of language is not communication but probably internal dialogue.

My issues with Millikan are slightly different. Despite the focus on lineages and proper functions, Millikan's view is detached from work in the biological sciences. Although she discusses coordination, she does not expand beyond individual interactions. It is also not clear how the formal structures charac-terised by formal grammars map onto the kind of patterns Millikan is referring to. The problem is that Millikan's conventions govern individual linguistic objects which are passed down and reproduced based on their specific survival values. These are determined by their individual profiles, not their structural or relational properties in the first place. This situation is inevitable given the focus on organism-level features. However, it is necessary to ascend to a higher level of analysis in order to find the linguistic patterns that correspond to the structures of the previous sections.

5.2 Unbanishing the 'Linguistic Community'

In this section, I want to *naturalise* the concept of the linguistic community. Mainstream linguistics has long banished the idea from relevance within the competence model of language. The idea of an external environment of speakers linguistically interacting in sometimes imperfect ways was con-sidered a 'theory of everything' (Chomsky 2000a) and as such a scientific nonstarter. Conventions, regularities and patterns among speakers within such a community, although favoured by some philosophers, have thus not received due theoretical investigation within the philosophy of linguistics. This latter perspective meant that the kind of naturalism often touted in generative linguistics focused on the individual level of an idealised competent speaker of a language. In other words, the real linguistic patterns discussed above map onto mental mechanisms of individual speakers at some higher level of abstraction (from brain-structures). Methodology in linguistics has, however, developed in different directions. It is now common to the point of commonplace to see large-scale corpus studies, cross-linguistic analysis, fMRI results, and a host of performance data feature in mainstream theoretical linguistics journals such as *Language, the Journal of Linguistics, Theoretical linguistics, Linguistics and Philosophy*, and *Lingua*. This shift indicates that the patterns being sought often outstrip the individual competence level and seek structural relations within the larger linguistic community.

Nevertheless, even if this is the trend, Chomskyans could still be correct that the idealisation of a linguistic community is infeasible and unworkable. I aim to show that by taking an approach analogous to systems biology, this idealisation can do exactly the work required of it. Thus, taking both the internalism of biolinguistics and the public externalism of Millikan's view into consideration, I advocate a systems biological approach to the material nature of language, *systems biolinguistics* if you will.

Systems biology is a holistic approach to the life sciences. It is an extremely collaborative interdisciplinary field which includes biology, computer science, physics, engineering, and mathematics. Whereas the nexus of traditional biology might have been individual organisms, cells, plant life etc. systems biology abstracts away from these to home in on their complex interactions with the environment. There are a number of specific sub-disciplines of this larger field, such as metagenomics or the study of diverse microbial communities.

Like many theoretical offshoots, systems biology started with critical reflection on the limitations of both standard microbiology with its focus on microbes such as viruses and bacteria and mainstream biology with its focus on individual macroorganisms such as plants and animals. For instance, classical concepts such as multicellurity are ill-defined on the entity-based accounts since they fail to capture the multi-cellular nature of symbiotic organisms like lichens which exhibit interdependent existence. Cellular cooperation, competition, communication, and certain developmental processes require a broader perspective than the object-oriented accounts can provide. Some have put forward the claim that microbial communities can be considered multicellular organisms themselves (O'Malley and Dupré 2007).

Dupré and O'Malley (2007) survey the literature on metagenomics or environmental genomics which "consists of the genome-based analysis of entire communities of complexly interacting organisms in diverse ecological contexts" (835). In this field, microorganisms are not placed in isolated artificial settings but rather assumed to be essentially coupled with their environments and interactions with other organisms. A proper investigation of biodiversity seems to require the analysis of metagenomes or large amounts of DNA collection within the environment. One additional reason for this shift is that evolution seems to require a larger perspective of this kind. As they state:

> Conceptually, metagonomics implies that the communal gene pool is evo-lutionarily important and that genetic material can fruitfully be thought of as the community resources for a superorganism or metaorganism, rather than the exclusive property of individual organisms. (Dupré and O'Malley 2007: 838)

On this view, one might consider human bodies to be complex symbiotic systems composed partly of human cells, viruses, the bacteria hosted by prokaryotes and so on. But this perspective is also too limited. Systems biology assumes that there is no non-arbitrary distinction to be had between an individual organism and its environmental conditions. No clear 'self' versus 'other' is discernible. The immune system is a clear case where the human host and the prokaryote communities form one complex system which benefits the organisms (Kitano and Oda 2006a, b). Dupré and O'Malley use these considerations and more to suggest an ontological shift is necessary and/or present in biology, one that moves from entities or organisms to processes and systems as the basic ontological categories. There is no useful concept of a static genome-organism correspondence as "[g]enomes, cells, and ecosystems are in constant interactive flux: subtly different in every iteration, but similar enough to constitute a distinctive process" (Dupré and O'Malley 2007: 841).[6]

Systems biology conceives of biological entities at the systemic level, not only as individual components, but interacting systems, processes and their emer-gent properties. In this way, Millikan was correct that language is like a dance in which coordination between partners plays a major role. What she left out was the interaction between the dancers and the dance hall, the other dancers at the party and human microbiota who call us home while we sweat and salsa.

In order to accommodate the analysis of big data, the complex inter-organism interactions and their environments, statistical and network approaches have become prominent. Thus, biological systems are usually represented as dynamic networks which form complex sets of binary interactions or relations between different entities and their contexts.

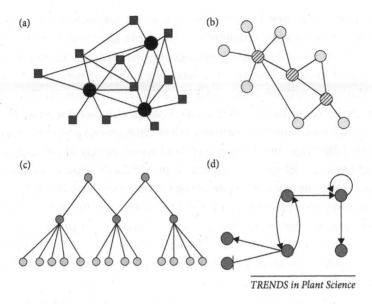

TRENDS in Plant Science

Fig. 5.1 Plant System Networks (from Yuan et al. 2008: 166).

Graph theory has been a very useful tool in the representation of biological networks. The vertices represent different biological entities such as proteins and genes in biological networks, and edges convey information about the links or interactions between the nodes. The links can be weighted or assigned quantitative values to encode various properties of interest, either topological or otherwise. More complex networks of networks can model the interaction between systems themselves (Gao et al. 2014).[7] These can take the form of *trees* or *forests*. See Figure 5.1 for different kinds of networks used on plant systems biology (no pun intended). Some networks model correlations across multiple conditions (a), while others (b) model sets of molecular interactions (or 'interactomes'), (c) shows hierarchical regulatory networks of genes with another way of modelling this shown in (d) by means of graphs that resemble finite-state automata.

Similarly, group theory has also been used in systems biology to great effect. Systems and networks admit the possibility of symmetries and graph automorphisms in rather straightforward ways but Rietman et al. (2011) review other work on symmetry breaking in genetics and cell cycle processes. Once we ascend to the level of systems, structural relations become more apparent. And the mappings between the kinds of structures we identified in the previous sections and the material mode, in this case the biological realm, are only possible beyond the individual organism level.

The novel suggestion I am making here is that real linguistic patterns are to be found in linguistic communities defined in similar ways to biological systems. In other words, natural languages are identified with the systems interpretation of linguistic communities involving complex interactions between individuals, internal structures, and the external environment. Thus, these communities comprise the internal rule systems of individual members and their externalised outputs at the material level. In essence, I propose a marriage between I-language and E-languages (and some aspects of 4E approaches to cognition, see Chapter 9), especially in Millikan's sense of conventions governing the communicative interactions between speakers. The full account also requires an appreciation of the constitutive effects of the linguistic environment on the production and consumption of language. This could involve socio-linguistic data and instances of cognitive coupling as in the case of the ubiquitous use of spell-checkers, dictionaries, and online translators like Google Translate.

Recently, Richard (2019) has proposed that we consider linguistic meanings in a similar biologically analogous manner. He argues that the transmission, exchange and change involved in linguistic meaning is biologically 'species-like', i.e., meanings are species. Importantly, he considers meaning to be "a dynamic, population-level phenomenon" (Richard, 2019: 4) and not on the individual level. Specifically he argues that word meanings are best understood "when we think of them as being like those segments of population lineages that we label species" (Richard, 2019: 3). His overarching project aims to rescue Quine's claims about analyticity without resorting to meaning scepticism. What's interesting about the account, from our perspective, is that it involves treating meanings as transmissible elements of linguistic communities viewed as biological systems in which the environment plays a key role.

Similarly, Miller (2021b) expands on the species connection in his account of the ontology of words. He rejects the standard essentialism about species (which dates back to Aristotle) in favour of a homeostatic property clusters view in which "the presence of some properties tends to favour the presence of other properties within the cluster" (Miller 2021b: 20). The clusters guarantee the causal powers reminiscent of kinds and these combined with underlying homeostatic mechanisms (which in turn cause the clustering properties) account for both species and words membership. The view argues for the necessity of word-kinds and then defines them in terms of property clusters. The mechanisms responsible for the clusters are both the internal-cognitive and external-social. In this way our views dovetail quite significantly. The difference is that I link the mechanisms to biological systems broadly defined.

Importantly for our purposes, there are two concepts of 'system' in systems biology. They differ in terms of ontological commitment. As O'Malley and Dupré (2005: 1271) state:

> The first account is given by scientists who find it useful for various reasons (including access to funding) to refer to the interconnected phenomena that they study as 'systems'. The second definition comes from scientists who insist that systems principles are imperative to the successful development of systems biology. We could call the first group 'pragmatic systems biologists' and the second 'systems theoretic biologists'.

The pragmatic approach dominates in the field. However, some systems biologists insist that such an approach offers little philosophical insight. Taking systems to be some collection or conglomeration of parts misses aspects of interconnection, emergent structures and symbiosis. The alternative, one I endorse here, is that "[s]ystems are taken to constitute a fundamental ontological category" (O'Malley and Dupré, 2005: 1271). How this category is constituted will depend on the type of biological system at hand.[8]

Notice the immediate improvement of this view upon pluralist ontologies. Language is not to be identified with hybrid ontological objects, part abstract or abstractish, part mental and somehow also physical in some ordinary object sense. Rather formal structures represent languages which are patterns found in complex biological systems created by a feedback network of conventions, linguistic competence and environmental interactions. These are *real patterns* in the sense of Dennett and Ladyman and Ross.[9]

Moreover, the view takes into consideration what Ludlow and Davidson identified as microlanguages and passing theories respectively, discussed in Chapter 3. Languages are not static but constantly in flux, dynamical systems. And Chomsky's worries concerning E-languages and their unsuitability for scientific inquiry is ill-founded in this case. Systems biology is extremely amenable to scientific investigation from statistical analysis, to group theoretic and graph-theoretic characterisation. Formal grammars characterise parts of these systems, and symmetries capture other aspects of their intra and interstructural relations (as we'll see in Chapter 7). In addition, the view tracks what Ladyman and Ross claim about the relationship between formal and material modes in terms of networks of mappings between levels. Most systems biology accounts make use of network analysis (including neural networks) as we have seen above. One such network that has already been exploited within the philosophy of language and evolutionary linguistics to a significant degree is signalling networks. A prominent example is Skyrms (2010) who, following Lewis (1969) (and in the spirit of this chapter so far) models propositional or semantic content as a special case of informational content thereby reintroducing information theory to philosophers of language and linguistics in terms of the emergence of linguistic communication and/or semantic meaning. He shows how signalling behaviour in humans and other animals (i.e. signs with content) naturally emerged as a result of biological evolution, learning, and certain adaptive processes.

In a similar vein, the work of Simon Kirby and his colleagues show that structure too naturally emerges from the complex system of innate signalling and communication. They consider the approach "a new way of thinking about the role of cultural transmission in an explanatory biolinguistics" (Kirby 2013: 460). Kirby focuses on the idea that language is an adaptive system.

> The evolutionary approach to this challenge [explaining why language has the structural features it does and not others] is one that attempts to explain universal properties of language as arising from the complex adaptive systems that underpin it. (Kirby 2013: 460)

Kirby too embraces complex systems analysis and designs his models so as to capture the essence of numerosity, i.e. not only the role of individual elements in emergent structure but also their interactions. 'Iterated learning models' in computational language evolution research aim to explain how complex syntactic structure, such as discrete infinity, is generated by creating highly simplistic models involving generational simulations of populations with no language to begin with (see Brighton and Kirby 2001). But importantly, Kirby and his colleagues see themselves as in some ways starting from a very different perspective to the Minimalist method of Chomskyans. Specifically, they claim that "rather than abstract away details about population structure or patterns of interaction, computational modellers will typically retain these complexities" (Kirby 2013: 461). Thus, the biolinguistic perspective is stretched to include population level dynamics but with the focus on language evolution *via* the emergence of phenomena like innate signalling and the role of iterated learning.

As previously mentioned, these systems also reject object-oriented accounts. Biological systems and hence linguistic systems modelled on them can be measured at various levels with distinct ontological commitments. At the level of systems, individuals disappear and structural relations dominate. Homing in on other nodes of the networks can illuminate different ontological aspects of the system at different scales. Put more strongly, "from this perspective *there are no biological objects (as metaphysically robust entities)*. All there is are biological structures, inter-related in various ways and causally informed" (French 2011: 172). Thus, *contra* the linguists above, language cannot be a biological entity on this view, without embracing global anti-realism, that is.

Thus, the advantages of this alternative foundation for biolinguistics is four-fold.

i. Systems biology tracks dynamics better than static individualist accounts (even those involving states).
ii. Systems are amenable to scientific investigation *via* group theory, graph theory and various network analyses.
iii. Systems biolinguistics rejects object-oriented accounts.
iv. Given [i.], [ii.], and [iii.], the position is more naturalistic and reflective of linguistic theory as a biological enterprise.

I hope to have shown that given the above, this proposal goes some way in fulfilling the promise of biolinguistics, which the generative programme has adopted to varying degrees of success.[10] Before I conclude this sketch, I want to show how this understanding of linguistics can shape or *naturalise* another central topic of biolinguistics, i.e., the idea of an I-language or steady state of the language faculty.

The term I-language is meant to capture the idea of a mature state achieved by a language learner after the primary linguistic data (PLD) has set various parametric settings of the innate UG capacity (Chomsky 1986). Unlike the alleged externalised or socio-political concept of a language like English or kiSwahili spoke in a particular community, I-langauges were supposed to be more scientifically tractable. However, a common criticism of this picture is that it produces a static view of language and ignores various dynamic aspects of the system. This is because that steady state or I-language is identified with a narrow concept of syntactic or the computational component of the faculty of language (Hauser et al. 2002). Where systems biolinguistics can assist is by reinterpreting this steady state of a language learner as a dynamic equilibrium in which "a system is said to be in 'dynamic equilibrium' or 'steady state' if some aspect of its behaviour or state does not change significantly over time" (Ladyman and Wiesner 2020: 72). In biological systems, this state is related to the concept of *homeostasis*. Homeostasis is in turn related to feedback from the environment (e.g. linguistic interlocutors in your community) and robustness of structure. Notice that the proposal here is not merely about nomenclature. Homeostasis is intimately linked to the environment. It is not a completely isolated internal system or UG only reflecting some sort of activation by external stimulus. Language is then not an internal component of a human mind or brain but a complex steady state of a biological system or linguistic community attained by intricate calibration with the linguistic (and non-linguistic) environments.

5.3 A Note on Acquisition

Metaphysicians often aim to describe some aspect of the furniture of the universe without much regard for how we learn how to sit on the couches they describe. In the philosophy of linguistics and theoretical linguistics in general, the issue of language acquisition cannot be so easily ignored. Whatever ontological picture of language a theorist puts forward, they need to contend with the question of how it is that young children acquire this entity, state or system (and with such ease!).

Platonists, for instance, face serious difficulty in accounting for how young infants begin the process of learning a nonspatio-temporal object. In general, acquisition accounts are split between more empiricist leaning ones and their rationalist alternatives. In fact, acquisition has often been the issue that makes or breaks a linguistic framework. Generative linguistics has long used the 'poverty of stimulus' (POV) argument to advocate for its brand of the innateness hypothesis. The argument goes something like this: young children are able to learn and master their native languages with phenomenal ease during their early acquisition period. This is not based on being exposed to copious amounts of data since they are never in such a position. Therefore, they must come prewired with a kind of language or 'mentalese' which gets activated by the specific linguistic environment into which they are born.[11] The P&P model is one way of explaining this process.

One worry might be that the complex structural picture I have painted in this chapter blocks the appeal of an empiricist theory of language acquisition. The more structure is assumed, the more rationalist accounts which are willing to impute as much structure as is needed innately seem like the only games in town. This might indeed have been true before the advent of Deep Learning in AI in the past decade or so. With this upgrade on the connectionist cognitive architectures, we have viable mega-empiricist machinery which offers the possibility of dealing with the kind of structuralist view of language proffered here. Let us start with some basics.

Deep learning is the most widely used and powerful computational tool in AI currently. Applications range from natural language tasks such as machine translation (e.g., Google Translate) and speech recognition (e.g., Amazon's Alexa), to image recognition, beating the world's best Go players, and medical analysis. Deep learning itself is based on neural network architectures similar to earlier connectionist models and that of Parallel Distributed Processing (PDP) (Rumelhart and McClelland 1986). Basically, a neural network is made up of nodes and links "intended to model the behavior of neurons and synapses at some level of abstraction" (Buckner 2019: 2). Of course, this is just an analogy. Deep learning is a part of machine learning which is in turn part of AI not neurobiology.

A neural network takes an input (e.g., text, numbers, or images), then processes it through a series of layers, in order to create an output in terms of either classification or prediction. Each layer contains a set of nodes which hold information and transmit signals to nodes in other layers. If there are more than two layers, the network is considered 'deep'. The links are weighted which either inhibits or activates a node. When the sum of incoming weights exceeds some designated threshold (depending on the function used: sigmoid, concatenate, softmax, max pooling, and loss), then the given node becomes active. Different architectures are used for different types of activities: Convolutional Neural Networks (CNN), are used in computer visual, e.g., for processing images, and Long Short Term Memory networks (LSTM), a type of Recurrent Neural Network (RNN) which are designed for processing sequences, are often used for linguistic tasks. The network learns a function based on a training set. Training consists of providing inputs and expected outputs, so the system can learn how an input should be processed. Sometimes this is done via supervised learning (i.e., labelling the data). The system is then tested with unseen inputs, to see if it is able to process these accurately. The structure of a network is abstractly show in Figure 5.2.

What neural networks are especially good at is picking up patterns hidden in complex sets of data. They do this by surveying sometimes hundreds of thousands of connections all the while self-correcting *via* a process called 'back-propagation'. The result is a hyper-empiricist framework for capturing the real patterns of complex systems in reality. Neural nets are exceptional at prediction tasks, sometimes in inexplicable ways (cf. the black box problem or epistemic opacity discussed Marcus (2018), Sullivan (2019), Nefdt (2020a)).

The point is that if language learners have access to anything like a neural network in their task of negotiating information from the environment, interactions with others and prewired dispositions, then they can in fact

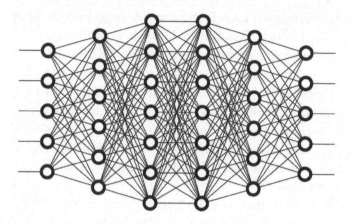

Fig. 5.2 DNN.

acquire a language defined in the structural terms presented here. This is of course not much more than an analogy. But given the empiricist nature of neural networks and the network structure of biological systems, it seems like a good one.[12]

Can a language (or LRP) in terms of a biological system of the likes I have been describing even be learned or acquired? There is some indication that it can be. Above, I followed systems biology in claiming that the immune system is a classic example of biological system. Here the work of Nobel Laureate Niels Jerne (1985) is particularly relevant. Using an analogy of generative grammar, he argued that the immune system acquires its antibody strategies in a manner similar to language acquisition.

> [I]nstead of producing an ad hoc response every time an antigen enters the organism, nature has provided humans with abundant repertoire of different types of antibodies. Some of them may never play an actual role if they never encounter a disease they can block, but some are already assembled in our body in order to allow the immune response to operate quickly. (Moro 2016: 19)

In fact, many contemporary systems biologists take the nature of the immune system to require their analysis. As Zak and Aderem (2009: 266) state: "[t]he complex interactions within the innate immune system that result in effective host defense under normal conditions and inflammatory disease when perturbed can only be dissected in a comprehensive way by systems biology approaches". Similarly, Dhillon et al. (2020: 2) claim that immunology requires a systems biology approach for "understanding the immune response to vaccination, infection and diseases, since these involve complex interactions between a large number of genetic, epigenetic, physiological and environmental factors". The immune system is a complex of multiple internal and external processes and organisms cooperating functionally within an environment to produce immune strategies for dealing with pathogens. It has innate aspects as well as specific systematic responses to features of the context. In fact, biologists discuss specific pattern recognition receptors (PRRs) in the innate immune system (see Zak and Aderem 2009). It is not implausible that our innate linguistic capacities or the language faculty contains such pattern recognising receptors. There is a lot work on pattern recognition in the Natural Language Processing (NLP) (Kocaleva et al. 2016) literature as well as the general cognition literature (Pi et al. 2008, Youguo 2007, Mattson 2014).

In terms of the external component of language specifically, Pereplyotchik (2017) makes a compelling case that Chomskyans owe us an account of both the 'normative-teleological' terminology involving correctness standards common in the acquisition literature and a characterisation of 'target' or 'ambient' language—the language in the linguistic environment of the language learner—without recourse to public E-languages. He doubts that such a theory would be possible. His own suggestion involves a measure of deviation of the child's idiolect from the grammar of the ambient language assuming a goal of fluid communicative interaction with members of the community. In Section 4.3, I motivated the tool of finite state transducers as mechanisms for comparing linguistic patterns. In the deep learning case, back-propagation performs the task of reevaluation and adjustment based on target values and weighting which might be analogous to what a language learner is doing.

At base, the idea is that we are pattern recognition machines and languages are complex (non-redundant) real patterns. We do come 'prewired' with some mechanisms, but the rest of language acquisition involves searching for patterns within complex systems and networks of interrelated interpersonal and environmental features, adjusting our internal states in accordance with the linguistic environment. Thus, it has been argued in this chapter that a systems biological approach can be adapted for capturing the structures and corresponding real patterns behind linguistic reality. Of course, it is not the only option compatible with the real pattern analysis proffered here. But I do believe it is a fruitful new avenue, if naturalism and the biological analogy are both to be maintained.

6

A Case Study

Words and SLEs

The task of the present chapter is to apply some of the ontological concepts from the previous chapter to the resurgent debate on the metaphysics of words. Nowhere in the philosophy of language and linguistics is the role of the type-token distinction so pronounced as in the debate over the ontology of words. The central insight I will be advocating can be described as follows:

> CENTRAL INSIGHT V: WORDS, PHRASES, AND RULES ARE ON A STRUCTURAL CONTINUUM WHERE THE ROLES THEY PLAY IN OVERARCHING LINGUISTIC STRUCTURES SERVE AS THEIR PRIMARY ONTOLOGICAL STATUS.

The chapter is broken up into three initial parts. The first briefly introduces the problem of word individuation. I then offer a naive or intuitive answer based on folk linguistics. Next, I follow contemporary cognitive linguistics and Jackendoff's parallel architecture in rejecting this simple picture before mounting a philosophical account based on the structuralist insights of the previous chapter and Szabó's (1999) representationalist treatment. The latter I use to gain traction on the real patternhood of the view. There are a number of extant views not discussed in this chapter. The main reason for this restricted scope is that my aim is mostly application in this chapter. For a comparison of some of these ideas with extant ideas on the market see my (2019b).

6.1 The Naive Picture and Three Naturalistic Desiderata

Words are an essential part of our understanding of linguistic practice and behaviour. They are often associated with the conceptual apparatus and ground syntactic theories (for instance as terminal nodes in syntactic trees). But what exactly are words? The answer is surprisingly hard to pin down. Are they physical tokens of abstract linguistic types as the type-token distinction would

Language, Science, and Structure. Ryan M. Nefdt, Oxford University Press.
© Oxford University Press 2023.
DOI: 10.1093/oso/9780197653098.003.0006

have it or are they individual objects each with a *sui generis* ontology like people (Kaplan 1990, 2011)?

In this chapter, I approach the question of the ontology of words in a more circumscribed manner. Following the dialectic in Chapter 4 (and Santana (2016)), I will mostly be interested in the question of which concept(s) of words is useful for their precise study, in terms of linguistics and the philosophy of language.[1] In other words, I will delve into the ontology of words in a manner directly informed by their scientific study as *per* the naturalistic approach of the book. In this vein, again, I propose three desiderata for evaluating different metaphysical proposals, namely (1) learnability, (2) metaphysical consistency, and (3) correspondence with current linguistic theory.

One might at first glance be tempted to consider words to simply be individual sounds and/or inscriptions which particular communities come to use in particular ways. These sounds come to have phonetic, syntactic and semantic properties. For example, the word *dog* is pronounced [dɒg], it is a noun which can be pluralised to form *dogs* and generally refers to the canine species frequently adopted as pets in households across the world.

An initial reason to worry about this simple nominalist picture is that words are often pronounced in very diverse ways, and the same sound can represent different words (words of different categories or with different meanings). *Dog* can be pronounced in various ways in terms of various dialects, the American pronunciation is different from the one above [dɔg] (and there are further variations in a particular New York accent [dwɔrg]). *Dog* can be a verb as in *Susan was dogged by persistent nightmares*. Semantically, *dog* can be used to refer to the class of animal mentioned above or a sexually promiscuous male. Even if we abstract over alternative meanings, alternative syntactic categories and alternative pronunciations, we still don't have a clear path between the noise and the word.[2] Furthermore, if words are identified with sounds we produce, then how do we explain that different languages use the same sounds for different things? Spanish speakers use ['bu.ro] for 'donkey', while Italian speakers use ['bu.ro] for 'butter', i.e. so-called false cognates.

A related issue concerns word individuation. How many words are there in the following sentence?

1. *A dog is a dog, is a dog.*

Eight or three? Do multiple instances of the word *dog* count as three or one word or does the indefinite article *a* count as one or three words? On the simple picture, there are three distinct words (sounds) in the above sentence since words are identified purely in terms of their phonetic profiles. In another sense, however, (1) is an eight word sentence with a main clause marked by the second *is* and a subordinate clause preceding it.[3]

In light of these difficulties, I propose three means of constraining the debate on the ontology of words, to which the naive picture fails to adhere. Proceeding in accordance with these desiderata does not amount to a knock-down argument of views which fail to conform to them, rather I argue that any adequate account of the ontology of words which aims to be useful for their scientific study should adhere to the following constraints or related ones.

Desideratum 1. An adequate account of words needs to be informed by an explanation of how we acquire words.

Desideratum 2. An adequate account of words should involve a naturalised linguistic ontology.

Desideratum 3. An adequate account of words needs to correspond to contemporary linguistic analyses of phrases, clauses, and sentences in which they figure, i.e., it needs to conform to a scientific treatment of natural language *simpliciter*.

It is plain to see that the naive picture fails all of the above conditions. In terms of (1), if a child were forced to learn the individuation of words exclusively based on physical properties such as their sound profiles then they would not be able to easily grasp words with the same sounds but distinct meanings or uses. Yet polysemy effects are commonly witnessed in early childhood language learning. "Some of the most highly frequent and earliest learned English verbs, like *put, make, get, go,* and *do*, are also among those with the largest number of senses" (Clark, 1996). According to Theakston et al. (2002) children as young as two freely understand and use many polysemous words, often with very little confusion. This suggests that they are appreciating more than just phonetic variation.[4]

In terms of the second desideratum or (2), the naive theory seems to imply a certain nominalism about words. Thus, on the face of it, it appears to commit us to a physicalist picture involving sounds, inscriptions and the like. This, however, produces questions as to how different instances or pronunciations of particular words all correspond to identical word-types. In other words, if what determines a word's identity is a confluence of real or rather apparent phonetic properties, then it is not clear what makes two tokens instances of the same word when they sound different. Additionally, the individuation problem above threatens to undermine a naive nominalist picture like this one.

In essence, this desideratum is a specialised case of the idea that the proposed view should fit with our best scientific theories on any subject matter whatsoever. Suppose the proposed account of words did not fit with our best theory of, say, photosynthesis? Presumably we would have to revise either the proposed theory of words or our current theory of photosynthesis.[5] In many ways, the failure of the naive picture on words is not surprising given that our folk physics (biology, psychology) is unlikely to map onto our best scientific theories. That is to say, there is a natural disjunction between the manifest image and the scientific one.

Lastly, this initial picture suggests that words are transparent or to be found in nature like rocks or trees. They wear their properties on their sleeves, so to speak. However, some contemporary linguists postulate constructs that could be interpreted as implying unpronounced invisible words (such as deleted copies, the understood second-person subject of imperatives, etc.). Furthermore, we need an account of words which explains their incorporation in larger linguistic structures such as clauses and sentences. Therefore, if we are to adhere to **desideratum 3**, we need a different sort of view to the naive one.

According to linguistic analysis, words are not all that apparent in general, especially phonetically. As Poole (2002: 24) notes:

> [I]f we were to look at a sound spectograph of [an English sentence], a visual representation of the sound energy created when [it] is produced, you would see that there are no breaks in the sound energy corresponding to where the word breaks are. It's all one continuous flow.

This is part of the reason why learning a second language can be so difficult: word boundaries are not easy to spot. Thus, although words might be physical things in part, they are not to be identified with apparent physical properties. In traditional metaphysics, it would seem we need a theory of words which accounts for how they are categorised into types which in turn can be learned by the speaker of the language through the features of their tokens.

6.2 Constructions and Constraints

Not all linguistic theorists take lexical items to be distinct from other kinds of SLEs such as morphemes, phrases, rules or sentences. For example, Jackendoff (2018) argues for collapsing the distinction between words and grammatical rules (pertaining to sentences, for instance) as both qualify equally as "pieces of structure" such that "[n]ot only is there no clear place to draw the line between words and rules [hierarchical tree structures that pertain to larger linguistic units], there is no need to" (104). For him, there is a cline between words and more rule-like structures. He suggests that model-theoretic grammars might offer a better formalism for capturing this continuum.

> The Declarative Strategy (some versions of which are called "model-theoretic" by Pullum, 2013) defines the repertoire of possible linguistic representations in terms of a set of well-formedness conditions on structure. (Jackendoff 2018: 101)

In essence, using constraints on well-formedness instead of generating expressions based on rules, allows words and sentences to be treated analogously in terms of the grammar.[6] This idea is not novel. It has been a seed in Jackendoff's own theory of grammar for decades and can be traced to work on constructions in construction grammar.

Over the years, Ray Jackendoff has challenged mainstream linguistic theory is a number of ways. His Parallel Architecture (PA) (Jackendoff 2002) has been especially critical of the autonmy of syntax. Not only does he not adhere to the idea that syntax is the sole generative engine of the language faculty but he also advocates separate but parallel generative components in terms of phonology, morphology, and semantics. What links these components are interface principles. He describes the project as follows:

> The alternative to be pursued here is that language comprises a number of independent combinatorial systems, which are aligned with each other by means of a collection of interface systems. Syntax is among the combinatorial systems, but far from the only one. (Jackendoff 2002: 111)

Interesting though this framework undoubtedly is (see Burten-Roberts and Poole 2006; Nefdt 2016 for more), we are not directly interested in the PA itself here, but rather its treatment of words which draws inspiration from cognitive linguistics, in particular construction grammar (see Chapter 8 for more on the PA).

Cognitive linguistics is a broad research programme which covers a number of distinct frameworks. The general idea involves bringing the study of language closer to the scientific results and theories of cognitive-neuroscience. Aspects of generative syntax such as the autonomy of syntax and domain-specificity (or modularity) are thus often eschewed in a majority of these accounts. The field itself emerged in the 1970s as a response to the formalist and computationalist approach of generative grammar. Its theoretical inspiration drew from the Gestalt theory, non-modular views in the philosophy of mind, and work on concepts such as prototype theory among other things. The linguistic framework fits well into the 'second generation cognitive science' with its focus on embodiment and situated cognition as opposed to the more abstract mentalist tradition of formal linguistics (Sinha 2010). We'll cover this territory more thoroughly in Chapter 9.

Construction grammarians reject what they describe as 'distributional analysis' which rests on a 'building block model of grammar'. One particularly strong advocate of this alternative linguistics, William Croft, states that view as:

> Grammar is seen as being made up of minimal units (words or morphemes) belonging to grammatical categories, and constructions are defined as structured combinations of these units. The purpose of the distributional method, therefore, is to identify the grammatical categories that are the building blocks, and the units that belong to those categories. (Croft 2013: 3)

However, Croft claims that distributional patterns do not match within and across languages. This causes tension for the building block model according to him since specific language constructions seem to define distinct categories (e.g., Absolutive-Ergative is not identical to Subject-Object etc.). The radical step is to jettison universal categories entirely and acknowledge that the building blocks differ for each language and perhaps more interesting for each construction of individual languages.

What exactly are constructions? Well, these are usually taken to be pairings of form and meaning such that there are "slots" for saturation of individual constructions as in the following (adapted from Hoffman and Trousdale 2013):

a. idiom construction: e.g., *X take Y for granted*—'X doesn't value Y'
b. comparative construction: e.g., *John is taller than you*—'X is more Adj than Y'
c. resultative construction: e.g., *She rocks the baby to sleep* [X V Y Z]—'X causes Y to become Z by V-ing'[7]

This situation leaves us with a contrary account to the standard framework of generative linguistics. Constructions are complex entities taken to be basic units of grammar. For instance, generative grammar under Minimalism operates on the opposite assumption that basic units are words or lexemes and *via* rules, more complex entities are generated. Goldberg (2015) mounts a version of this argument against the principle of compositionality (assumed in formal semantics) which adheres the the building block model of grammar. Constructions, she argues, are ubiquitous in natural language and a failure to acknowledge this can led to untoward consequences (such as infinite verb senses if the projection principle of generative grammar is true, see Goldberg (2015: 6) for more).

However, the acceptance of semi-fixed constructions in grammar, as assumed by construction grammar, is not tantamount to rejecting the building block model by itself. For instance, one could assume that grammatical constructions are on a scale from completely decomposable (or flexible) to less so (e.g., from novel structures to fixed idioms). Jackendoff (2002, 2018) assumes such a *structural* continuum. All that is required, following classical cognitivist treatments of constructions (such as Fillmore, Kay and O'Connor 1988) is to distinguish between syntactic configurations in terms of 'schematicity', with the more schematic involving more syntactic and morphological variation. Idioms are less schematic on this picture since many of their parts do not undergo morphological inflection etc., such as *kick the bucket* in which only the verb is grammatically flexible. A syntactic rule is interpreted as entirely schematic on this account. Furthermore, it's constructions all the way down. Not only syntactic rules but also lexical items can be reinterpreted as constructions (albeit less schematic ones). This perspective treats all SLEs on par with one another instead of assuming linguistic theory involves distinct ontological categories. Jackendoff (2017) explains the concept in the following manner:

> I propose that there is a lot to be gained by abandoning the strict separation of lexicon and grammar and thinking of rules of grammar as part of the lexicon. The argument goes by a sort of slippery slope: We can find a succession of items stored in the lexicon that are progressively more and more rule-like. As we move through this progression, there is less and less reason to distinguish the items we find from things that everyone considers to be rules. Moreover, when the bottom of the slippery slope is reached, it turns out not to be as absurd as it might initially appear. (192)[8]

Furthermore, as we have noted above, Jackendoff (2018) convincingly argues that model-theoretic syntax or constraint-based grammars are preferable to generative grammars since they can capture the structural continuum of linguistic items to better effect. I use the term 'structural continuum' to avoid the implication that the differences between the items in the unified ontological category is a matter of degree only. For example, one plausible way of capturing the structural continuum might be Pietroski's (2018) internalist semantics. His account says that words, phrases, and rules are all instructions: words are simple instructions to fetch a concept, phrases are complex instructions to fetch a concept, rules are instructions to combine concepts. It's not a matter of degree, for him. Similarly, the continuum I am suggesting is not one in which sentences are larger words but that words, phrases and sentences are all rule-like in related ways. We don't need to take a stand on the issue of model-theoretic grammars here either, especially given the multiple models perspective we will advocate in the next chapter. In the concluding parts of this chapter, I apply the naturalistic ontology argued for in Chapter 4 to SLEs and thereby attempt to move the immovable classical debate on the ontology of words.

Before this, a note on acquisition or **desideratum 1**. In the previous chapter, I suggested that acquisition can accept empiricist principles despite involving a central role for structure. In a sense, the continuum presented here makes the job of an acquisition story slightly easier. Instead of a separate mechanism for the acquisition of words and other SLEs, linguists only need to posit one unified learning mechanism. Construction grammar approaches are generally considered to be a species of usage-based acquisition theories. These approaches allow for innate constraints or learning biases but emphasise the network interactions of different components of the individual-environment relationship as advocated in the systems biology analogy of the previous chapter. Diessel (2013: 348) states the acquisition picture as "a dynamic system of conventionalized form-function units (i.e., constructions) that children acquire based on domain-general learning mechanisms such as analogy, entrenchment, and automatization." Not only would this learning approach reduce the ontological difference between words and rule-like structures for learning but it will reduce the gulf between linguistic learning mechanisms and general purpose ones. Surely, a plus for parsimony! However, despite the natural fit or appeal between this kind of view and the structural ontology, the latter maintains its compatibility with standard generative approaches to language acquisition.

6.3 A Structural Approach to Linguistic Entities

Recall, from Chapter 4.2, that mathematical structuralism is the view that mathematics is the science of abstract structures, where the objects in these structures are defined purely with relation to each other and/or the overarching superstructure. Mathematical objects, thus, have no internal or individual natures in themselves on this view. Different structuralist proposals differ on what they take these structures to be, eliminative or object-preserving, and what the background theory looks like, modal or set-theoretic, first-order or higher-order, etc. For instance, the particular brand of structuralism for which I advocated earlier, called *ante rem* or noneliminative structuralism (Shapiro 1997; Resnik 1997), does not eschew individual mathematical objects *per se* but finds a place for them within a structuralist ontology, based on an analogy with *universals* in the history of metaphysics.

For present purposes, there are two features of structuralism on which I want to focus briefly in order to advocate for the analogous analysis of linguistic items such as words *qua* positions in structures. The first is how individual objects are treated in the theory (I will follow *ante rem* structuralism here) and the second relates to how structures themselves differ ontologically from Platonic objects.

As we've seen, according to one particular interpretation of structuralism, there are two ways of conceiving of objects *via* the resources of natural language. In the one case, we treat positions as offices or roles, which are multiply realisable in terms of entities. For instance, some uses of *President* or *bishop* are examples of these cases. They do not denote individual objects as in *The President has the right to overrule the senate* or *The bishop can only more diagonally.* Shapiro calls this 'places-as-offices'. There is another sense of the term in which we treat positions not as the offices or roles they occupy but as genuine singular terms denoting objects. Examples are sentences such as *The President had lunch with the Dalai Lama today* or *The bishop ate the King at d7.* This is the 'places-are-objects' perspective. *Ante rem* structuralism takes this latter concept as primary. Of course, as Shapiro notes "[w]hat is an office from one perspective is an object -and a potential officeholder- from another" (1997: 11).

In arithmetic or number theory we take numbers to be objects, but in set theory they are offices. Consider the number 2, "[i]n one system, [finite von Neumann ordinals] $\{\emptyset, \{\emptyset\}\}$ occupies the 2 place, and in the other [Zermelo numerals] $\{\{\emptyset\}\}$ occupies that place" (Shapiro 1997: 11). In either case, the numeral 2 is a name picking out an object *qua* position in a structure and statements involving the numeral are true or false. Nevertheless, in neither case are we committed to an individually existing number in the Platonic sense, just the structures.

Thus, we have two distinct ways to treat words, as offices or placeholders in linguistic structures and as *bona fide* objects in themselves. The former works well for words without overt semantic or even phonological structure such as the complementizer *that*, elliptical words or PRO in linguistic theory. PRO is the postulated subject of nonfinite clauses such as *John wants Mary* [PRO] *to visit*. But it is highly tendentious (and not necessary) to claim that PRO is a word, in fact it might be better considered an empty phrasal position containing no words. There are many theoretical alternatives available on the market.

There is one aspect of the structuralist ontology which I think is helpful in distinguishing them from ordinary objects as Linsky and Zalta (Chapter 1) urge for Platonised Naturalism. As we have seen, for the Platonist, abstract objects are analogous to ordinary physical ones in that they are ontologically independently of one another. My laptop doesn't rely on my desk for its existence. Mathematical objects, on the other hand, have no such independent existence according to structuralists. These objects *qua* positions in structures depend on other positions for their very existence and on the structures as a whole. Linnebo (2008a) distinguishes between two notions of *dependence*.

ODO Each object in D [domain of some mathematical structure] depends on every other object in D. (67)

ODS Each mathematical object depends on the structure to which it belongs. (68)

The difference between ODO and ODS is that the former states that the objects in a structure depend on other objects such as some natural numbers depending on other natural numbers.[9] While the latter adds that the existence of one object in a structure ensures that the structure itself exists or rather is ensured by the existence of the structure as a whole. Another way to think of ODS is that structures are ontologically prior to positions (some structuralists like to say that there would be structures even if there were no objects fulfilling the various roles).

The first [*ante rem* structuralism] takes structures, and their places, to exist independently of whether there are any systems of objects that exemplify

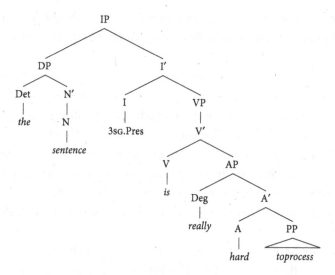

Fig. 6.1 *The sentence is really hard to process.*

them. The natural-number structure, the real-number structure, the set-theoretic hierarchy, and so forth, all exist whether or not there are systems of objects structured that way. (Shapiro 1997: 9)

Importantly, if types are on the level of offices (in the sense discussed above) the analogy with physical objects is dropped, since these offices are not complete (do not have determinate truth values for all properties), do not have hidden natures and are certainly not sparse (offices can be created *ad infinitum* independently of entities discovered to fill those positions).

6.3.1 Linguistic Structures, Again

That generative linguistics is a structural enterprise should not be surprising to its practitioners. Linguists are generally concerned with hierarchical systems, trees with nodes representing different categories ultimately terminating in lexical units or words as in Figure 6.1.

The above structure evinces the two kinds of objects discussed in the previous section. Under the *I'* category (pronounced I-bar), we have a placeholder for the inflectional category (third-person singular, present) which like real words has features such as number and tense. Under the *N'* the word *sentence* is treated as a bona fide object or terminal element in the structure. Linguistic structures also display both kinds of dependence relation. Items such as the former depend on the overarching structure (the tree) for their existence (ODS) while words or terminals depend on other words and positions (ODO and ODS) for their existence. In general, syntacticians are not interested in words or clauses in isolation but how they work together to form phrases and ultimately sentences of the language. Thus, no word is an island (although some do have *island effects*).

There is a distinctly Wittgensteinian flavour to this suggestion.

> We see that what we call "sentence" and "language" has not the formal unity that I imagined, but is the family of structures more or less related to one another [...] The philosophy of logic speaks of sentences and words in exactly the sense in which we speak of them in ordinary life [...] We are talking about the spatial and temporal phenomenon of language, not about some non-spatial, non-temporal phantasm. But we talk about it as we do about pieces in chess when we are stating the rules of the game, not describing their physical properties. The question 'What is a word really?' is analogous to 'What is a piece in chess?' (PI §108)

A chess piece has physical dimensions or reality (even if it is bits of code in a computer programme) but in order to understand what a particular piece is one needs to appreciate its role or position in the structure of a game of chess. In terms of my **desideratum 3**, the above picture dovetails with actual linguistic practice. Consider a determiner like *the* or *an* (DET in the figure 6.1 above). On most syntactic accounts, it is a structurally designated linguistic position in a hierarchical structure or tree and any word or object (sometimes nothing as in the case of null determiners) can satisfy the position. And whatever is in that position is a determiner since it is identified by distribution (not meaning or sound). Of course, this is a means of identifying the type and not every instance. For example, the syntactic distribution of 'five' and 'six' might be the same but that these are nonetheless different words. Semantics, therefore, acts as another layer in individuating words at the material level. But these too are semantic structures linked to the full real patternhood of the item (see Chapter 8). As we've seen in this chapter, the postulation of covert material is also usually supported by structural reasoning in linguistics, i.e., something must be there since this structure requires it or it stands in a structural relation to something else.[10]

The theory so far provides a view which is informed by contemporary linguistics, i.e., fits **desideratum 3**, and in so doing provides a broader plausible framework for the learnability of words or **desideratum 1**. In the next section, I will expand on the ontological question (**desideratum 2**) by further synthesising this view with Szabó's representational account of types and tokens.

6.3.2 Words as Positions-in-Real Patterns

Let us return to the issue of **desideratum 2** or the ontological question. Here, I follow Szabó's representationalist picture. However, as he admits, ontology posed a problem for his view. I will show that the more nuanced picture offered here of the ontology of words does not share this problem.

Szabó pitches his framework in direct contrast to frameworks which endorse 'the instantiation view' or "[a] type T is instantiated by its tokens, and it is in virtue of this that empirical information about a token of T can play a role in justifying our knowledge about T" (1999: 147; cf. Bromberger's (1989) *Platonic Relationship Principle*). Szabó's characterisation goes further to insinuate that most views which incorporate types as special kinds of sets or patterns with their tokens exemplifying a certain kind of (projectible) similarity are implicitly committed to the instantiation view.[11]

We have already seen some problems with the traditional type-token distinction. Generally, types are identified by features of the tokens, either the way they sound or are spelled (what Kaplan (1990) calls 'phonographic' features). However, these features are generally unreliable sources for getting at the nature of types. Szabó follows Kaplan, Wetzel, and others in pointing out that 'recognitional criteria' (based on the 'physical appearance' of tokens) are often (although not always) unhelpful in the pursuit of type identification.

> Categorizing tokens like this would make types linguistically widely hetero-
> geneous, in a way that would imperil the reliability of inductive inferences
> from tokens to types. Even if phonological and orthographic criteria are
> acceptable for some purposes, they are unacceptable for explaining our
> linguistic knowledge of types. (Szabó 1999: 148)

The point is that the forms or physical appearance of words, sentences, and other tokens are part of a motley assortment of tools we use for the identification of types. Proponents of the instantiation view tend to neglect this fact. It is important to note that Szabó's claim is not that the instantiation view is untenable, quite the contrary, he believes it in principle to be 'simple and plausible' but he offers an alternative which aims to better capture the nature of the relationship between types and tokens. It is on to this proposal or 'representational view' that we now move.

A good starting point for the appreciation of the difference between the representational view and its Platonic rival, is an insight which dates back to Aristotle. 'Types are nothing more than abstract particulars'. The central idea of this claim is that types are incapable of instantiation or they cannot have instances. In Ancient Greek syntax, attaching the definite article to any noun (or infinitive, if the neuter article is used) results in an abstract version or concept pertaining to the referent of that word. For instance, from 'good things' (one word in Greek) we derive 'the good', from the infinitive 'to do wrong' to 'the wrong' or simply 'injustice'. In English, singular terms often fulfil this role. "We talk about the first line of Gray's *Elegy*, the last words of Goethe, of the fourth letter of the Hebrew alphabet" (Szabó 1999: 152). In these cases, words, infinitives and singular terms are being *used* to fulfil a certain role, namely a representational one.

The sort of reasoning employed in the previous paragraph indicates a functional approach to the role of tokens and their relationship to types. The function of a token is to represent a type and it is commonground that they play this role in the language. In this way, tokens are representations which stand 'proxy' for what they represent. There is an element of arbitrariness to this relation. Unlike the idea that tokens are instances of an overarching type, which suggests a more intrinsic picture of the role of tokens, on this view tokens are related to the class of objects which have Grice's non-natural meaning (Grice 1957).

The story goes on to specify exactly what kind of representation relation tokens bear to their types. Tokens represent their referents indirectly: they represent their type, which in turn represents the referent. By way of example, Szabó writes that "[t]he English word-type 'horse' represents horses and so do all its tokens. But unlike the word-type, those tokens represent only *indirectly*: they represent the word-type 'horse', which in turn represents horses" (1999: 150). Thus, this view encompasses two stages from tokens to the referents of the types, we learn about the nature of types not only by looking at the features of the tokens but also by knowing their roles as representations. To use a variant of Szabó's example, if we want to know about a particular species of animal, we use not only reference to that animal but also reference to representations of it (as found in books or pictures). Tokens are then in this sense very much like models in scientific discourse. They act as intermediaries between the types and the kinds of things that types represent. The important difference between this view and the instantiation account is that instances require additional ontological connections with their types that representations do not. This is why a sign in a spoken speech act in Cockney English, Morse code, and a written word can all represent the same word type without issue. To find the set of shared features through which they instantiate that type is a more daunting task than to understand their roles as representations.[12]

This view allows for a lot of flexibility which in turn can explain more uses of tokens. The problem, Szabó acknowledges, is that many of the reasons for and against this proposal can seem to lead to an irresolvable conclusion (as with many of the original debates concerning universals). What is a natural explanation for a representationalist might be unnatural for an instantiation-alist. In light of this, Szabó has a specific argument based on inverted spectrum arguments, tailored to bring out the advantages of the former position. We will not consider that here (see, Nefdt (2019b)).

In terms of the dialectic of this chapter, his account does better than its rival on the first desideratum of learnability. If word-types are representative in the aforementioned way, it explains why various media all work to establish a word-concept relationship in young children. A picture of a horse, the word *horse* said out loud and a sign for horse all work just as well to pick out horses, even if they have never encountered one. Words play functional roles and these roles can be learned from experience. As yet, no abstract objects are invoked, nor does the view distort linguistic analyses of words or sentences.

However, in terms of **desideratum 2** or the ontological question or the larger issue of naturalism, there is a profound worry within this proposal. The worry is that despite all the best efforts to distinguish the representational view from the instantiation view, when confronted with ontological questions of what words (or word-types) are, the former allows for a version of Platonism to resurface like a dormant virus. Furthermore, the ontological package with which this view comes is supposed to be purely physicalist. Types or those things which are represented by tokens are not supposed to be ontologically occult, they are not Platonic Forms. Certainly, there is nothing inconsistent about supposing that the *representata* are physical in some sense but exactly what this sense is is not an easy matter to specify. This is where Szabó's account becomes (self) admittedly 'tentative and speculative'. I think that it is at this point that the representational view can be buttressed by the structuralist interpretation of linguistics offered here.

Recall that structuralism was developed as an ontological alternative to Platonism. It is at this point that the view might be of use to the representational view of tokens and their types. Szabó has a distinct account of what types are not, i.e., Platonic Forms. His view of what they are, ontologically speaking, is more vague. He holds that types are created, not discovered, and thus that they have starting points in time. The problem with this claim is that it requires a certain creative power to be attributed to representation, since it presupposes that there must have been a first token of any given type. Without ever-existing Platonic types of words, this claim is hard to maintain.

In terms of ontology, Szabó makes a few ingenious observations but also concedes too much in others. As for the former observations, he proffers a more nuanced picture of representation such that it can occur before and independently of the existence of a *representatum*. On the intuitive copy-model of representation, word-types must exist before they can be represented by tokens, as an original must exist prior to its copies. This latter model, however, cannot account for some cases of representation such as the one below.

> Most of the work of an architect consists in producing representations—floor plans, drawings, models, detailed descriptions—of buildings that do not yet exist. Once we abandon the copy-model of representation, there is no difficulty here. The representation view can coherently maintain that the first tokens of a new word-type are much like the drawings of the architect: they represent something that does not yet exist. (Szabó 1999: 162)

I think that there is something to this idea but it is too quick in its current form. An adherent of the copy-model could object that the architect does not produce representations of non-existent buildings but rather uses existing representations to assemble novel structures. In this way, the creativity of representation resembles the creativity of language use in which structures can be combined to create novel structures. At the sentential level, Szabó's idea is sound. There are types of sentences determined by the rules of language, much like there are types of buildings determined by the laws of design and physics, that have yet to be tokened. The architect's blueprints represent something more abstract than a particular building, they represent a building structure-type. The point here is that attempting to describe representations *sans* the *representatum* is problematic if construed in a piecemeal manner but with relation to systems of structures it is more plausible.[13] Nor do the structures themselves need to exist *in toto* for the process to work. As the rules of the language develop and evolve, so do the systems of rules which represent the yet-to-be represented structures of that language.

The copy-model also incorporates an independence claim or the claim that the thing represented must exist independently of its representations. The necessity of this condition on representation is also challenged by Szabó by means of the national borders case. There are various ways of representing a national border (fences, walls, xenophobic attitudes, etc). If these ways of representing are eliminated, so too are the borders, Szabó argues. Thus, the border does not exist independently of the objects which represent it. Indeed, a national border seems like a good candidate for a quasi-concrete object.

Nevertheless, consider whether this reasoning applies to objects like the equator. There are ways in which we can represent the equator, imaginary lines across the circumference of the planet (or real lines drawn on representations of the planet), perhaps a long physical tube across the surface of the earth, etc. However, if these representations are destroyed the equator remains. One could think of a national border analogously.[14] Of course, destroy the planet and with it go the equator and all national borders but that is beside the point for now. The problem with the example is that one could ask how the representations of the borders are set up in the first place if there were no independently existing templates from which to work. For this reason, I think that these examples fall into the camp of bad prospects for a 'quick resolution' which Szabó dismisses earlier in his paper. Worse still, they force him to concede that certain linguistic types "cannot exist untokened" (Szabó 1999: 162). Furthermore, this possibility then produces a distinction in kind between the complex expression types (e.g., clauses and sentences) and the simple expression types (e.g., morphemes and word roots) of which these types are composed. He offers a Kaplanian suggestion for how the histories of word-types might have transpired to lend credence to this new distinction.[15]

On a more structuralist account the concessions made above, that tokens can represent in the absence of types and that words and sentences are representationally distinct, are not strictly necessary. In terms of the latter, following Kaplan on word histories and the like might commit us to the same error which Bromberger (2011) points out, namely that we are starting from a narrow conception of word-change over time as opposed to a more empirically sound conception of language (or structure) change over time. Bromberger urged two separate points. The first was that a proper account of word-type needs to account for the structural elements of words, i.e., 'that words function as constituents of phrases and sentences'. He took this to be the essence of words, that they function in larger structures 'whatever their intrinsic perceptual and referential features'. The point might be overstated and I think that a proper account of words should consider both representational and structural components of words (*qua* objects and offices respectively). The second point was that the focus on initial dubbings and individual word change common across the literature is a red herring. In other words, Kaplan's attempt to anthropomorphise words is a non-starter. In fact, the sociological perspective is a better place to start than the individualistic one.

> Normally, single words do not change in isolation, but whole families of words that share features change together as certain shared constituent features get replaced in shared phonological environments. (Bromberger 2011: 497)

He goes on to cite examples from the Great Vowel Shift in English to the Valley Girls Rise in North American dialects. I think the point generalises beyond phonology. Furthermore, I think that this line of objection might be damning for Kaplan's (and Hawthorne and Lepore's) view but not so for Szabó's. The representational view buttressed with a structuralist account of words as positions-in-patterns offers a route to accommodating both of Bromberger's worries. Firstly, structural elements of words are readily accounted for in this framework since it is at base a structuralist view of linguistic objects. Secondly, any changes are naturally pitched at the level of structures and substructures through which natural languages are characterised. The view I am pushing is thus in tune with linguistic literature on language change which is often described as law-like (as Bromberger asserts) and integrated, in that changes have structural effects and do not generally occur in isolation with individual histories. It is therefore not guilty of the linguistic *façons de parler* Bromberger so strongly opposes for the ontology of words in line with **desideratum 3**. Nor do we have to drive a wedge between our ontological treatment of words-types and more complex expression types such as sentences. When words change, they change their phonetics, syntax and semantics within a larger biological linguistic system. The changes have structural ripple effects that are lost when the debate on the ontology of words is conducted solely in terms of an individualist, object-oriented framework.

A caveat before a conclusion. I'm not saying that words are to be individuated exclusively by means of their syntactic profiles. Although I do think this is a neglected element within the contemporary (and erstwhile) debate, it's not the entire story. Phonological, semantic, and even pragmatic structures matter too. At this point, syntax illustrates structuralism more clearly for the sake of the case study. In later parts of the book (Chapter 8), I will add a layer of semantic structuralism to the full account.

6.3.3 Conclusion

This chapter has been an exercise in application. I used the naturalistic framework developed in earlier chapters to attempt an answer to questions in a debate in the philosophy of language by extending it into the philosophy of linguistics. Many contemporary accounts such as Miller (2021a), Irmak (2019), and Gasparri (2019), although meritorious in other ways, do not place the role of linguistics above other metaphysical considerations.[16]

I have put forward the claim that words are on a structural continuum or cline with other SLEs and perhaps syntactic rules in general. When they act, they act as places-in-concrete structures or real patterns as we have been calling them in Chapter 4. Where I expanded upon my previous account was by saying more about the identification of the real patterns in terms of the representation relation of Szabó (1999).

7

Structural Realism and the Science of Linguistics

7.1 The Aim and Scope

This chapter moves away from direct ontological questions which occupied the pages of the book so far to issues in the philosophy of science. Drawing from some of the previous insights, I aim to offer a structural realist interpretation of the science of linguistics, specifically focusing on syntax.

There are two central issues that will be dealt with in this chapter. The first, (1) is the intratheoretical problem of theory change in generative linguistics and (2) the second is an intertheoretical issue of multiple and competing grammar formalisms. Both of these problems threaten the success of formal linguistics. I hope to show that structural realism offers an account which not only respects the methodological (and ontological) nature of the discipline but also solutions to both aforementioned issues.[12]

Thus, I argue that linguistics as a science essentially faces the problem of pessimistic meta-induction, albeit at a much faster rate than the more established sciences such as physics. In addition, I claim that the focus on the ontology of linguistic objects, such as words, phrases, sentences, etc. belies the formal nature of the field which is at base a structural undertaking. Both of these claims, I argue, lead to the interpretation of linguistics in terms of *ontic* structural realism in the philosophy of science (Ladyman 1998, French 2006). To be *realist* in this sense is to accept the existence of linguistic structures (not individual objects) defined internally through the operations of the grammars and what remains relatively stable across various theoretical shifts in the generative paradigm, from Standard Theory (1957–1980) to the Minimalist Program (1995–present), are the structures so defined.

Ontic structural realism, however, comes with different settings. Set on high, it involves the claim that 'all there is, is structure' (what Psillos (2006) calls 'pure structuralism' or van Fraasen (2006) calls 'radical structuralism'). The burden of proof for such a claim would surely involve more than just reflection on the special sciences. In fact, one might worry that such claims amount to

Language, Science, and Structure. Ryan M. Nefdt, Oxford University Press.
© Oxford University Press 2023.
DOI: 10.1093/oso/9780197653098.003.0007

a statement that the world is mathematical in structure. For our purposes, we do not need this strong view or even the claim that linguistics is purely structural. All we need to capture is the idea that structure is *prior* to objects (and properties) in linguistics. This priority claim is considerably weaker than the global or pure one above. It also leaves room for the possibility that the structural analysis isn't exhaustive of theory change or model comparison. I leave the issue open as to whether other non-structural considerations can also play a role in the explanation of these phenomena. Thus, the central insight can be summed up as follows:

CENTRAL INSIGHT VI: THE PRIMARY VEHICLE OF THEORY CHANGE WITHIN GENERATIVE LINGUISTICS AND THEORY COMPARISON ACROSS FRAMEWORKS IS STRUCTURE.

The chapter is separated into four parts. In the first part, I focus on some important theoretical shifts which the generative linguistic tradition has undergone since its inception in the late 1950s. For instance, the move from rewriting systems with transformations to X-bar representation (Chomsky 1970) with theta roles to the current single movement operator *Merge* contained only by constraints. Despite appearances, I hope to show that the general structure of these representations have remained relatively constant. In the second part, I discuss both realism and structural realism in the philosophy of science more generally and why the latter might serve as an illuminating foundation for linguistics, assuming the former. Linguistics here is interpreted structurally without recourse to the independent existence of individual objects in that structure (along the lines of Shapiro 1997 for mathematics). In other words, there are no phrases, clauses or sentences outside of the overarching linguistic structure described by the grammar. In the next section, I link talk of invariance in physics *via* graph theory to invariance in linguistic theory as a tool to identifying structure. Finally, I discuss alternative grammatical frameworks ('the problem of multiple grammars'), especially dependency grammar and constraint-based approaches, and show how the current analysis can incorporate structural overlap between them and the models of generative linguistics.

7.1.1 The Challenge

The main motivation for structural realism has come from work in fundamental physics. Worrall (1989) famously reflected on the retention of Fresnel's equations within Maxwell's to defend a version of epistemic structural realism calling it the 'best of both worlds' strategy for avoiding the pessimistic metainduction and antirealism respectively. Ladyman (1998) bases his argument against the latter on the hole argument in general relativity and the status of particles in quantum theory, French and Ladyman (2003) prescribed the more ontologically committing alternative in which "structure is ontologically basic" (46) citing examples from Maxwell's electromagnetic theory of light to reflections on Poincaré. Even objectors, such as Psillos (1999) and Saatsi (2017) focus their critiques on the mathematical and physical theories. As Ross (2008) states "[o]ntic structural realism (OSR) is crucially motivated by empirical discoveries of fundamental physics" (732). Although Ladyman and Ross proffer 'rainforest realism' as a unification of fundamental physics with the special sciences (which take measurements from a restricted segment of the universe as confirmation). They describe the *challenge* in following way:

> Special sciences seem more clearly committed to self-subsistent individuals, and to causal relations holding among them, than do the fundamental theories of physics. If fundamental relationships in physics are described by symmetric mathematical relations (as they may or may not be), the same surely cannot be said of the most important generalizations in biology or economics. (Ladyman and Ross, 2007: 196)

Given the overwhelming emphasis on physics within the realism debate in general, one might worry that the *best* arguments for structural realism come from physics. Thus, this situation presents us with a particular challenge, that of showing that the view is applicable and perhaps even exemplified by scientific theories outside of physics. The specific challenge is stated in as French (2011, 2014) as that of identifying structures without the kinds of laws and equations of physics, especially in fields where contingency plans a larger role such as biology.

There is a growing literature in the philosophy of science on extending the remit of structural realism to disciplines other than those in the natural sciences. For instance, Kincaid (2008) discusses the view with relation to the social sciences (as we will see below), Ross (2008) to economics specifically, Ladyman (2011) to the history of chemistry, as mentioned French (2011, 2014) to the biological sciences, and Yan and Hricko (2017) to cognitive neuroscience (albeit a more localised version of SR). I myself, in Nefdt (2019c) and Nefdt (2021), developed aspects of this interpretation for the science of language. Here I will go a step further to establish the applicability of structural realism within generative linguistics and beyond it.

7.2 Linguistic Theory Change

The history of science bears witness to a number of radical theory changes from Newtonian physics to Relativistic, from Euclidean geometry to Riemannian as a characterisation of physical space, from phlogiston theory to Lavoiser's oxygen theory, among countless others. In the course of such changes, one might easily dismiss the old theory as simply false. Laudan (1981) famously proposed that there might be a deeper issue at stake here, namely what has become known as pessimistic meta-induction (PMI). PMI can be defined as follows for present purposes.

PMI : If all (most) previous scientific theories have been shown to be false, then what reason do we have to believe in the truth of current theories?

The problem with radical theory change is that it causes serious tension for any realist theory of science, which wants to hold to the truth or approximate truth of current theories. Of course, false theories can be responsible for true ones through some sort of trial-and-error process. But the idea that our best current theories are of mere instrumental value for later truth is hard to accept.[3] Furthermore, at no point will certainty naturally force itself upon us, especially since success is not a guarantee of truth (e.g. classical mechanics is still a useful tool for modelling physical phenomena). PMI has an ontological component as well. When theories do change, they often propose distinct and incompatible entities in their respective ontologies. Consider the move from phlogiston theory to oxygen theory. In fact, the term 'phlogiston' has become synonymous with a theoretical term which does not refer to anything.[4] Essentially, the ontological status of the objects of the theories are rendered problematic when radical theory change occurs, which prompts a challenge again to the realist. "[I]f she can't establish the metaphysical status of the objects at the heart of her ontology, how can she adopt a realist attitude towards them?" (French 2011: 165).

Linguistics too has seen its fair share of radical shifts in theory and perspective over the past few decades. In fact, the early generative tradition of Chomsky (1957) had a more formal mathematical outlook. Drawing inspiration from the work of Emil Post on canonical production systems which are distinctively proof-theoretic devices in which symbols are manipulated via rules of inference in order to arrive at particular formulas (not wholly unlike natural deduction systems), linguistics approached language from a more syntactic perspective.[5] This was due in part to two assumptions, namely that (1) syntax is autonomous from semantics, phonology, etc. and (2) that syntax or the form of language is more amenable, than say semantic meaning, to precise mathematical elucidation. Mathematical models of this sort would be a key tool in early generative linguistic analysis. Chomsky (1957: 5) stated the formal position in the following way at the time.[6]

> Precisely constructed models for linguistic structure can play an important role, both positive and negative, in the process of discovery itself. By pushing a precise but inadequate formulation to an unacceptable conclusion, we can often expose the exact source of this inadequacy and, consequently, gain a deeper understanding of the linguistic data. More positively, a formalized theory may automatically provide solutions for many problems other than those for which it was explicitly designed.

He goes on to chastise linguists who are skeptical of formal methods. However, as we shall see, the course of linguistic theory saw a decrease in formalisation and an increased resistance to it (partly inspired by Chomsky's later views). In fact, a generative grammar in the early stages was expressly noncommittal on ontological questions. "Each such grammar is simply a description of a certain set of utterances, namely, those which it generates" (Chomsky 1957: 48). By the 1960s, grammars were reconceived as tools for revealing linguistic competence or the idealised mental states of language users. With mentalism, linguistics looked towards sciences such as psychology, physics, and biology for methodological guidance as opposed to logic and mathematics as it had before. As Cowie (1999: 167) states of the time after *Aspects* "[Chomsky] seemed also to have found a new methodology for the psychological study of language and created a new job description for linguists". The psychological interpretation of linguistic theory held sway until the 1990's when the 'biolinguistic' programme emerged as yet another new way of theorising about language.[7] The *Minimalist Program* (1995) pushed the field towards understanding language as a 'natural object' in which questions of its optimal design and evolution take centre-stage.[8]

Each new foundation distanced itself from the methodology of its predecessor, postulated different objects and advocated different ends. Thus, PMI takes on special significance for linguistics and an answer to the puzzles it presents become especially peremptory in this light. In the following sections, I will focus on some specific cases of the methodological changes which underlie the above picture.

7.2.1 From Phrase Structure to Phase Structure

In this section, I aim to provide a story of the mathematical formalisms employed in the service of an ever-changing landscape of theory in linguistics. Many of the theoretical postulates, such as 'deep structure vs surface structure', the modules of Government and Binding theory, domain specificity, optimality, I-language etc., of various generations of generative grammar are not explicitly dealt with in this chapter.

The early generative approach had a particular notion of a language and accompanying grammar at its core. On this view, a language L is modelled on a formal language which is a set of strings characterisable in terms of a grammar G or a rule-bound device responsible for generating well-formed formulas (i.e. grammatical expressions). In *LSLT*, Chomsky (1975: 5) writes of a language that it is "a set (in general infinite) of finite strings of symbols drawn from a finite "alphabet"". In the earlier Chomsky (1956), natural languages were shown to be beyond the scope of languages with production rules such as $A \rightarrow a$, $A \rightarrow aB$ or $A \rightarrow \varepsilon$ (ε is the empty string) such that $A, B \in NT$ and $a \in T$ (i.e. regular languages).[9] This result led to the advent of phrase-structure or context-free grammars with production rules of the following sort: either $S \rightarrow ab$ or $S \rightarrow aSb$ (read the arrow as 'replace with' or rewrite). These grammars can handle recursive structures and contain the regular languages as a proper subset. For many years, phrase-structure grammars were the standard way of describing linguistic phenomena. Essentially, phrase structure grammars are rewriting systems in which symbols are replaced with others such as $S \rightarrow NP, VP$ or $NP \rightarrow det, N'$. As Freidin notes "phrase structure rules are based on a top-down analysis where a sentence is divided into its major constituent parts and then these parts are further divided into constituents, and so on until we reach lexical items" (2012: 897). There are a number of equivalent means of representing the structure of sentences in this way. The most common is *via* hierarchical diagrams, shown below.

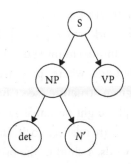

1.

Alternatively one can capture the same information as:

2. $[_S[_{NP}[_{det}][_{N'}]][_{VP}]]$

This basic structure, however, proved inadequate as a means of capturing the structure of passives and certain verbal auxiliary constructions, as shown originally in Postal (1964). Transformations were meant to buttress the phrase structure system in order to bridge this gap in explanation. Transformation rules operate on the output of the phrase structure rules and create a derived structure as in (3) below for passivization.

3. NP_1 V $NP_2 \rightarrow NP_2$ be-en (AUX) V NP_1

The combined expressive power of phrase structure and transformations proved very productive in characterising myriad linguistic structures. This productivity, with its increased complexity, however, came at a cost to learnability. "[I]f a linguistic theory is to be explanatorily adequate it must not merely describe the facts, but must do so in a way that *explains* how humans are able to learn languages" (Ludlow 2011: 15). The move to more generality led in part to the Extended Standard Theory and the X-bar schema.

Since the continued proliferation of transformations and phrase structure rules were considered to be cognitively unrealistic, linguistic structures needed more sparse mathematical representation. Although, as Bickerton (2014a: 24) states "'rule proliferation' and 'ordering paradoxes' were only two of a number of problems that led to the eventual replacement of the Standard Theory".[10]

There was also some theoretical push for more general structure from the Universal Grammar (UG) postulate assumed to be the natural linguistic endowment of every language user. UG needed to contain more general rule schemata in order to account for the diversity of constructions across the world's languages. This structural agenda dovetailed well with the Principles and Parameters (P&P) framework which posited that the architecture of the language faculty constituted a limited number universal principles constrained by individual parametric settings, where 'parameters' were roughly the set of possible variations of a given structure (see Chapter 2.3.1). The so-called Extended Projection Principle might be universal but certain languages can contain distinct parameters with relation to it (such as fulfilling it with a null determiner). In other words, a child in the process of acquiring her first language can 'set' the parameter based on the available linguistic environment in which she found herself, like flicking a switch. Furthermore, this kind of structural picture is represented well in the X-bar schema (Jackendoff 1977) which contains only three basic rules. There is (1) a specifier, (2) an adjunct, and (3) a complement rule. The idea is that the schema effectively treats endocentric projection as an axiom, which the previous phrase structure rules did not. 'Endocentric' here roughly means that one element (i.e., the head) of a constituent determines the function and nature of the whole. The X-bar schema, in other words, restricts the class of phrase markers available (this was part of Chomsky's (1970) original motivation at least).

The specifier rule is given below (where X is a head-variable and XP and YP are arbitrary phrasal categories determined by that head).

Specifier rule: $XP \rightarrow (Spec)X'$ or $XP \rightarrow X'(YP)$

Or equivalently:

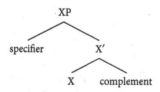

A vast amount of linguistic structure can be modelled by means of this formalism.[11] In fact, X-bar theory over-generates structural descriptions (which need to be reined in by various constraints). But the underlying idea is that our mental competence is more likely to contain generalised rule schemata such as those above than individual phrase structure rules and countless transformations for each natural language. In a sense, X-bar merely smooths over the individual hierarchical structures of before and homes in on a more abstract structural representation for language. As Poole (2002: 50) mentions:

[W]e discovered that your language faculty appears to structure phrases into three levels: the maximal projections or *XP* level, the intermediate *X′* level, and the head or *X°*.

These rules subsume the previous *ad hoc* phrase structure rules. Importantly, the representation, however, only allows for binary rules (unlike the possible n-ary branches of phrase structure trees). Freidin (2012) further claims that X-bar theory represented a shift from top-down to bottom up analysis, despite being formulated in a top down manner a decade into its inception. Here, the idea is that the rules stated above are projections from lexical items to syntactic category labels not the other way around.

Unfortunately, history has a way of repeating itself. Where in the previous instantiation of generative grammar, the proliferation of transformations became unwieldy, parameters would soon see a similar fate befall its fecundity. Briefly, UG was assumed to be extremely rich during this period, "the available devices must be rich and diverse enough to deal with the phenomena exhibited in the possible human languages" (Chomsky 1986a: 55). However, what was innate and what was learned or set by experience relied in part on a distinction between 'core' grammar and 'periphery', never explicitly provided by the theory (see Pullum (1983) and Culicover (2011) for discussion). Although, formally all previous transformations were reduced to the 'move alpha' operation, the multiplication of parameters took similar shape to its transformational predecessor. Newmeyer (1996: 64) describes this period as one of instability and confusion.

In the worst-case scenario, an investigation of the properties of hundreds of languages around the world deepens the amount of parametric variation postulated among languages, and the number of possible settings for each parameter could grow so large that the term 'parameter' could end up being nothing more than jargon for language-particular rule.

What's more is that these parameters seemed to force the violation of the binary requirement set by the X-bar formalism and with it the cognitive plausibility transiently acquired after the Standard Theory. There needed to be a better way of capturing the movement toward simplifying the grammatical representation and theory of natural language syntax. This and other theoretical motivations led to the Minimalist Program (1995) which pushed the new biolinguistic agenda and a call for further simplicity.

As mentioned in Section 7.2, the question of the evolution of language reset the agenda in theoretical linguistics at this time. The grammatical formalisms assumed to underlie the cognitive aspects of linguistic competence were forced to change with this new perspective, with the result that many of the advances made by the P&P and Government and Binding (1981) theories needed to be abandoned (according to Lappin et al. 2000).[12] Of course, abandonment is a strong claim. Many linguists consider GB to have been on the right track but too complex in its analysis while MP merely filters the structures to only involve the 'conceptually necessary' (again, in line with the structural realist interpretation I proffer below). The rationale was something of the following sort.

> Evolutionarily speaking, it is hard to explain the appearance of highly detailed, highly language-specific mental mechanisms. Conversely, it would be much easier to explain language's evolution in humans if it were composed of just a few very simple mechanisms. (Johnson 2015: 175)

The *Merge* operation represented the goal of reducing structure to these simple mechanisms. In the Standard and Extended theories, grammars followed the structures set by the proof theory in the early twentieth century (see above) which often resulted in grammars "of roughly the order of complexity of what is to be explained" (Chomsky 1995: 233). In the Minimalist Program, this apparatus was reduced to a simple set-theoretic operation which takes two syntactic objects and creates a labelled output of their composition (the label to be determined by the features of the objects thereby replacing the projection from heads of X-bar theory).[13] The formulation is given below:

7. $\text{Merge}(\alpha, \beta) = \{\gamma, \{\alpha, \beta\}\}$

Or again, equivalently:

$$\gamma$$
$$\overset{\mid}{\underset{\alpha \quad \beta}{\wedge}}$$

The above is an example of external set *merge* (where γ is a label projected from one of the elements). Internal *merge* accounts for recursive structures since it applies to its own output (as in if β is already contained in α). Consider the following sentence.

8. The superhero should fly gracefully.

In a bottom up fashion, *fly* and *gracefully* will *merge* to form a VP, thereafter this union will *merge* with the auxiliary *should* to form a TP or Tense Phrase. *Merge* will independently take *the* and *superhero* and create an NP which will *merge* to form the final TP to deliver (8) above (the T is the label projected for the entire syntactic object). Importantly for the proposal I will present, "[t]his last step merges two independent phrases in essentially the same way that generalized transformations operated in the earliest transformational grammars" (Freidin 2012: 911).[14] Thus, although the phrase structure rules had been replaced by the less complex *merge* operation with *phases*, which are cyclic stages applying to the innermost constituents of the entire process (Chomsky 2008), the structure is identical in the derivation.

Of course, unlike the top-down analysis of early generative grammar, *Merge* operates from lexical items in the opposite direction (*Merge* and the 'lexical array' constituting 'narrow syntax', see Langendoen 2003). As shown in the example above, it does apply to more complex units and their outputs. However, as Lobina (2017) cautions "talk of top-down and bottom-up derivations is clearly metaphorical" (84).[15] It might add something in appreciating the flavour of the computational process at hand, but often the overall structural picture is unchanged by such parlance.

Lastly, the notion of a *phase* is relevant here. A phase is created when the construction of a constituent *XP* is followed by access to the lexicon. This can occur when a lexical item can be inserted into a matrix *CP* (complementizer phrase) in cases in which earlier insertion, in an embedded CP, would have delayed movement. More importantly for our purposes, from the definition of a phase, we get the *Phase Impenetrability Condition* or the claim that if X is dominated by a complement of a phase *YP*, X cannot move out of *YP*.

Although phase theory was introduced in Chomsky (1998), one aspect of its structure predates this introduction by three decades, namely so-called 'island effects' (Chomsky 1964, Ross 1967). This is a massive topic in linguistics, so I will briefly focus on the Wh-island constraint and its similar treatment in early generative grammar and by means of phases in the more contemporary setting here. Consider the two sentences below:

9. Which book did Sarah say Mary liked?
10. * Which book did Sarah wonder whether Mary liked?

The above examples show a few things about the structure of Wh-movement. Movement itself is generally taken to be unbounded, but there are structures that can block it. For instance, (10) shows that Wh-movement can be blocked in embedded clauses containing *whether*. Both (9) and (10) show that movement happens in small steps (from CP to CP) since if it happened in a big step from the bottom of the tree in (9), then (10) should be licensed likewise.

Island effects were initially explained by means of the A-over-A principle or "if a rule ambiguously refers to A in a structure of the form of (i), the rule must apply to the higher, more inclusive, node A" (Chomsky 1964).

   i ...$[_A...[_A...]$
   ii 1. I won't read $[_{NP}$ the book on $[_{NP}$ syntax $]]$.
       2. *Syntax, I won't read the book on.
       3. The book on syntax, I won't read.

The embedded *NP* in (ii.1) is blocked from moving in (ii.2) by the principle (later subsumed under the Empty Category Principle or ECP). The island blocks the movement, where an 'island' is understood as a constituent that 'traps' items from moving out of them.

But this phenomenon can be explained in terms of phases as well.[16] A Wh-island arises when the SpecCP in the middle is already full. Since the Wh-word in the embedded clause cannot be moved into SpecCP, it gets trapped. The CP phase completes, and the higher interrogative *C* can no longer access the wh-word because it is inside of a finished phrase as in (11).

   11. *Which book$_i$ did Sarah think who$_j$ [who$_j$] wanted to read [which book$_i$]?

The explanatory strategy involves certain structural configurations which block the movement of items in embedded units or phrases. Another way of capturing this is that certain phases (CPs or vPs) do not allow Wh-movement to proceed through their specifiers (*Spec*). These phases are then the islands. There is a clear shift from the definition of islands in the A-over-A principle to their definition as phases *via* the Phase Impenetrability Condition in Minimalism. Despite this, the strategies for dealing with Wh-islands are similar from a structural point of view (as will be argued in the next section).

Let this serve as an account of some of the formal and theoretical changes of generative grammar over the 80 year period since its inception. Below, I will draw on the picture developed here to argue for the structural continuity of linguistics despite the theoretical shifts the overarching theory might have taken during this time.

7.3 Structural Realism in Generative Linguistics

The previous sections showed a theory in flux with each new stage seemingly jettisoning the achievements of the last. In such a scenario, the PMI seems especially problematic. Not only this, but as mentioned before, the situation in linguistics is unique since practitioners of each epoch of the theory can still be found working within the remit of their chosen formalism. In section 7.2, I described some of the theoretical shifts in the generative paradigm since the 1950s. In Section 7.2.1, I described the underlying mathematical formalisms utilised in service of the changing theory at each junction. In this section, I want to use a structural realist analysis of linguistics to show that despite the former, the structures of the latter remained relatively constant or at least commensurable.

What is structural realism? One way of thinking of it is as the 'best of both worlds' strategy for dealing with PMI. Realists, as we have seen, have trouble holding on to the objects of their theories once better theories come along. Anti-realists, on the other hand, have trouble accounting for the unparalleled predictive and explanatory success of theories (whose objects don't refer to objects in reality). Structural realism offers a conciliatory intermediary position between these choices. Ladyman (1998: 410) describes the position as follows.

> Rather we should adopt the structural realist emphasis on the mathematical or structural content of our theories. Since there is (says Worrall) retention of structure across theory change, structural realism both (a) avoids the force of the pessimistic meta-induction (by not committing us to belief in the theory's description of the furniture of the world), and (b) does not make the success of science [...] seem miraculous (by committing us to the claim that the theory's structure, over and above its empirical content, describes the world).

There are two versions of structural realism in the philosophy of science. The first, initially proposed by Worrall (1989), is epistemic in nature. The second, championed by Ladyman (1998), French and Ladyman (2003), is an ontological proposal. The former involves the idea that all we can *know* is structure, while the latter is a claim about all *there is*. In other words, what is preserved across theory change is a kind of structure posited by the underlying equations, laws, models or other mathematical representations of the theories. Part of the reason I opt here for ontic structural realism is that there is an ontological component to PMI as mentioned before. Thus, we are not only interested in what is communicated or epistemically accessible between different theories over time but what these theories say exists as well. Both versions agree on the existence of structures. Where they differ is on their respective treatments of objects.[17] Ontic structural realism takes an anti-realist stance here while the epistemic variety is agnostic. Thus, the ontological answer to PMI is therefore that if we cannot be realists about the objects of our scientific theories, we can be realists about the structures that they posit.[18]

From here, it is not hard to see what the argument of the present section is going to be, namely that different generations of generative grammar display structural continuity notwithstanding variation in theoretical commitment. The means by which we can appreciate this continuity is by considering features of the mathematical representations employed during the course of history which could affect my proposed analysis. Moss (2012: 534) has a similar idea when he discusses the contribution made by mathematical models to linguistic theory.

> [L]anguage comes to us without any evident structure. It is up to theoreticians to propose whatever structures they think are useful [...] Mathematical models are the primary way that scientists in any field get at structure.

In the previous section, I told a story about how the proof-theoretic grammars of the Standard Theory were transformed into X-bar representations which eventually led to the *Merge* operation in Minimalism. However, a remarkable fact about the structural descriptions generated by these various formalisms is that they share a number of essential features, namely, (1) expressively equivalent (also called 'weak generative capacity'), (2) they take a finite input and generate an infinite output, and (3) they can be represented hierarchically through tree structures. None of these latter properties are trivial. For instance, dependency grammars can be shown to be weakly equivalent to phrase structure grammars but are represented by means of flat structures. Model-theoretic grammars, such as Head-Driven Phrase Structure Grammar, are usually hierarchically represented and can generate the same sets of sentences but do not have any cardinality commitments. In other words, these features are *preserved under various transformations* of linguistic theory (a particular means of identifying structural identity, see next section). We will return to the issue of multiple grammars in the next penultimate section of this chapter.

Before I move on to a discussion of what structural properties could be and how to identify structures within linguistic theory, it is important to note that there were a number of formal shifts present in the transitions from transformational grammars to *Merge*. I have already mentioned the top-down to bottom-up change and argued that from a structural point of view, this is largely a metaphorical distinction. There is, however, another property of formal representations of syntax which also shifted from early to later generative grammar, namely from derivational approaches to representational or constraint-based ones. Simply put, derivational approaches follow the proof-theoretic model discussed earlier, where given a certain finite input and a certain set of rules, a particular structured output is generated. Constraint-based formalisms operate differently. Rather than 'deriving' an expression as output from a rule-bound grammar, these formalisms define certain conditions upon expressionhood or what counts as a grammatical sentence of the language.

Chomsky discusses this shift in thought in the following way:

> If the question is real, and subject to inquiry, then the [strong minimalist thesis] might turn out to be an even more radical break from the tradition than [the principles-and-parameters model] seemed to be. Not only does it abandon traditional conceptions of "rule of grammar" and "grammatical construction" that were carried over in some form into generative grammar, but it may also set the stage for asking novel questions that have no real counterpart in the earlier study of language. (Chomsky 2000b: 92)

Indeed, with the Minimalist agenda and the *Merge* operation, more constraint-based grammar formalisms were embraced and adopted. This latter approach contains a different idea of 'rule of grammar' and indeed 'grammar construction'. The formal difference can be understood in terms of how each type of formalism answers the so-called 'membership problem'. Decidability is an important aspect of formal language theory. Given a string w and a formal language $\mathcal{L}(G)$, there is a finite procedure for deciding whether $w \in \mathcal{L}(G)$, i.e., a Turing machine which outputs 'yes' or 'no' in finite time. In other words, a language $\mathcal{L}(G)$ is decidable if G is a decidable grammar. This is called the membership problem. What determines membership in a traditional proof-theoretic grammar is whether or not that string can be generated from the start symbol S and the production rules R. In other words, whether that string is recursively enumerable in that language (set of strings).[19] What determines membership in a constraint-based grammar is whether the expression fulfils the constraints set by the grammar (which are like axioms of the system). "An MTS [model-theoretic syntax] grammar does not recursively define a set of expressions; it merely states necessary conditions on the syntactic structures of individual expressions" (Pullum and Scholz, 2001: 19). As mentioned above, GPSG and HPSG are formalisms of the latter variety, while phrase structure grammars fall within the former camp.

The interesting fact for our purposes is that *Merge* and Minimalism represent the fruition of the gradual shift from derivational grammars to constraint-based ones. However, Chomsky (2000b) does not initially put much stock in this formal transition despite the strong statement quoted above. He considers the old derivational or "step-by-step procedure for constructing Exps' approach and the 'direct definition ... where E is an expression of L iff ... E ..., where ... - ... is some condition on E' approach to be 'mostly intertranslatable" (Chomsky 2000b: 99).[20] Here he holds these formalism-types to have few empirical differences, I will consider this thought in more detail in the next section.

From a mathematical point of view, the same formal languages and the structures of which they are composed are definable through both generative enumerative and model-theoretic means. Traditionally, the formal languages of the Chomsky Hierarchy were defined in terms of the kinds of grammars specified at the beginning of the previous section. However, there are other ways of demarcating the formal languages without recourse to generative grammars. For instance, they can be defined according to monadic second order logic in the model-theoretic way. Büchi (1960) showed that a set of strings forms a regular language if and only if it can be defined in the weak monadic second-order theory of the natural numbers with a successor. Thatcher and Wright (1968) then showed that context-free languages "were all and only the sets of strings forming the yield of sets of finite trees definable in the weak monadic second-order theory of multiple successors" (Rogers 1998: 1117).

The point is that the same structures can be characterised by means of proof-theoretic or model-theoretic techniques. Thus, the move from the former to the latter should not be seen as a hazard to the structural realist account of linguistic theory I am proffering here. In fact, in the next section I hope to show that this situation provides strong support for this particular analysis of the history and philosophy of linguistics.[21]

Lastly, the analysis suggested here dovetails naturally with other proposals to extend the purview of structural realism beyond physics and chemistry. For instance, Kincaid (2008) discusses the possibility of such an analysis for the social sciences. He argues that for structural realism to be successful vis-á-vis the social sciences, it needs to be shown that "social scientists talk about structures and not individuals" (Kincaid, 2008: 722) and that when such talk occurs "the individuals do not matter and the structure does" (724). In other words, social theories which emphasise 'roles' and 'relations' over and above the individuals occupying those roles or standing in those relations count in favour of a structural realist analysis. Kincaid offers three cases which meet the aforementioned condition, (1) general claims about social structure (e.g., organisations, classes, groups, etc.), (2) the cases of causal modelling (and a reinterpretation of the problem of 'underidentification'), and (3) equilibrium explanations (involving the relations between self-consistent variables).

Similarly to these cases in the social sciences, linguistics (especially syntax) provides examples of structure trumping individuals. There are a number of examples in syntax, the most stark of which is the positing of covert material or items based purely on structural considerations. Covert material in syntax refers to elements of the derivation that receive no phonological spell out. In other words, they are unpronounced items licensed only by the fact that the syntactic analysis requires a certain role to be played. Simple cases involve the EPP principle mentioned above (where a language can posit a null subject to fulfil the structural requirement) and DPs or determiner phrases which need not contain actual determiners (such as *a(n)*, *the*, *every* etc.). Another example is the PRO postulate in syntax (discussed in the previous chapter). This element is an entirely *null* noun phrase (or empty category) which means it too goes unpronounced phonologically. This analysis figures in infinitival constructions in which PRO is said to operate as the subject of infinitives, *Mary wanted John [PRO] to help her.* The behaviour of this structural element PRO is different from that of general anaphors, referring expressions, and pronouns, which means it gets its own category despite not being visible to surface syntax. The idea is that something needs to fill the role in order for the overall structure to work, and thus PRO is postulated.

Thus, in line with Kincaid (2008), linguistics can be shown to have cases (I would argue, many more than the social sciences) in which 'individuals do not matter' and structural considerations drive explanation. As he points out, there are general claims concerning structure, in our case phrases, X-Bar (as we've seen), trees, and operations on trees; specific cases of structural analyses such as negation and the general positing of covert structure; and even movement, an essential component of generative grammar across its time-slices, in which an item moves from one position in the tree to another, is not motivated by the individual nature of that item but the structural constraints on the grammaticality of the phrase or expression in which it is found.[22] Therefore, it would not be a stretch to consider linguistics, and syntax more so, to be a structural enterprise and thus amenable to a structural realist analysis.

Essentially, establishing that structural realism (whether epistemic or ontic) is a viable ontology for a series of theories requires two conditions to hold. The first is that they can be expressed structurally (in the sense of Kincaid). I have done so above for linguistics. The second is that their structures can be shown to be equivalent or isomorphic (or at least some related weaker structural relationship pertains). Section 7.2.1 made the case for the latter condition.

However, Kincaid's conditions might serve us well in motivating a general structural realist framework for a given science but it does not answer the question of what exactly is structurally preserved across specific theories. For this task, Ladyman's (2011) comparison between phlogiston theory and Lavoisier's oxygen theory is useful.

> Phlogiston theory subsumed the regularities in the phenomena above by categorizing them all as either phlogistication or dephlogistication reactions where these are inverse to each other. This is a prime example of a relation among the phenomena which is preserved in subsequent science even though the ontology of the theory is not; namely the inverse chemical reactions of reduction and oxygenation [...] The empirical success of the theory was retained in subsequent chemistry since the latter agrees that combustion, calcification and respiration are all the same kind of reaction, and that this kind of reaction has an inverse reaction, and there is a cycle between plants and animals such that animals change the properties of the air in one way and plants in the opposite way. (99)

Here he suggests that phlogiston theory meets a commitment of structural realism (both epistemic and ontic) in being a case of the "progressive and cumulative" nature of science and "the growth in our structural knowledge of the world goes beyond knowledge of empirical regularities" (Ladyman, 2011: 98). Similarly the trace theory of movement although replaced with the copy-theory retains this structural knowledge of how to account for movement (cf. island effects in Section 7.2.1). If we follow the analogy with phlogiston, neither phlogiston nor traces have reference to anything in the world but the structural strategies employed by the earlier theories were empirically successful to a certain extent and thus retained in the later ones.

The above case is a relatively clear example. Other cases are not as transparent. Consider again the move from phrase-structure grammars to the X-bar schema to *merge*. It is not obviously the case that the same structures are preserved across formalisms, at least not without additional stipulations. Phrase structure grammars, for instance, do not inherit their categories or function from their parts as is the case with X-bar theory. This property is called endocentricity (as we saw at the beginning of the chapter). In X-bar theory, a sentence (previously S - exocentric) is taken to be an Inflectional Phrase projected from the verb (endocentric). You can capture this property with *merge* but only by means of labels. Headed constructions (endocentric) can be and are represented in many phrase structure rules. However, they are not *essentially* endocentric. Rather linguists have traditionally restricted themselves to the endocentric formulations implicitly. Whereas the X-bar formalism makes this property explicit. Consider the rules for NPs, VPs, PPs below:

i $NP \rightarrow Det, N$
ii $VP \rightarrow V, NP$
iii $PP \rightarrow P, NP$

Besides the explicit endocentricity of X-bar theory, the formalism also showed that specific rules can be generalised to structures involving the categories of SPEC, Head, and Comp, across the board. In other words, all of the rules from (i) to (iii) (and many more) can be simply captured by either of the structure rules shown in Section 7.2.1 during the discussion of X-bar.

Thus, the 'progressive and cumulative' growth in our structural knowledge is based in the realisation of the generalisability of headed constructions and projection. A structural feature inherited by *Merge* (with labels) in an even more abstract manner.

Nevertheless, without a more precise notion of structure or structural property, the analysis only serves to illuminate structural similarity. The next section aims to make more precise the notion of structure at play and in general how structural comparisons can be achieved.

7.3.1 Structure and Invariance

The last aspect of this account of the scientific nature and history of generative linguistics will involve a brief detour into the ontology of structures again, this time from a different route. In so doing, I hope to suggest a particular path, in line with a proposal from Johnson (2015), for how linguists might identify the relevant structures of their science, especially with relation the PMI.

In terms of what a structure is, we've seen the most common set-theoretic definition found in the literature. "A structure S consists of (i) a non-empty set U of individuals (or objects), which form the domain of the structure, and (ii) a non-empty indexed set R (i.e. an ordered list) of relations on U, where R can also contain one-place relations" (Frigg and Votsis 2011: 228). Another term for such structures is 'abstract structures' which means that both the objects in their domain of U and the relations on R have no material content (i.e. they need not be interpreted). Although the set-theoretic notion is commonplace, it remains controversial. Landry (2007) convincingly argues that different contexts require different structures (Kincaid (2008) similarly argues for a case by case application of structural realism). Muller (2010) rejects both the set-theoretic and category-theoretic (Awodey 1996) account in favour of an entirely novel approach. And a number of others propose alternative frameworks such as the graph-theoretic approach of Leitgeb and Ladyman (2008) and Leitgeb (2020) (see note 27).

In Chapter 4.2, I briefly mentioned groups as canonical examples of mathematical structures. In fact the ubiquitous use of group theory in physics has prompted many philosophers of science to adopt structural interpretations of fundamental physics and natural science more generally (French 2014). One reason for this is that the theory makes precise notions of structural invariance and the symmetries it characterises. This makes it the case that in both relativistic and quantum physics, the significance of symmetries, invariance, and group theory is unassailable. So far, we have been defining group-theoretic notions set-theoretically. Here a 'symmetry' denotes some sort of transformation of a structure or object which leaves it unchanged in some relevant respect. The symmetry group, in group theory, is just the group of all transformations under which the object is *invariant*. French (1999, 2000, 2014) traces and examines the role of group theory in the development of quantum mechanics. Similarly, Bonolis (2004) describes the history of group theory in mathematics and physics. I won't go into those details here. Importantly, however, group theory offered physicists a way of describing certain structures without recourse to object-oriented ontologies.[23] Specifically, the history and transition from classical mechanics to quantum statistics emphasises the concept of symmetry which is associated with non-individuality.

There exist deep connections between group theory and graph theory. And since linguistic trees are essentially graphs, I want to briefly discuss both the formal relationship between groups and graphs as well as the possible ontological import of graph-theoretic structure.

The most well-known correspondences between the two structures are Cayley graphs. But there are a number of other proposals on the market. Before we say more, let us briefly describe some basics of graph theory.

A graph is a mathematical structure that is defined in terms of two components: (1) nodes or *vertices* and (2) *edges* connecting pairs of vertices (or ordered pairs of nodes). More formally, a *graph* is an ordered pair $G = <V, E>$ consisting of a nonempty set V (i.e., the vertices) and a set E (i.e., the edges) of two-element subsets of V. They generally come in two flavours: directed and undirected. Simply put, an undirected graph is graph is one in which all the edges are bidirectional. A directed graph or 'digraph' by contrast is one in which the edges point in a direction. Lastly, a graph is *connected* if there is a path from any vertex to any other vertex, where a *path* is a sequence of vertices using the edges (basically just like drawing a line from one node to another in the graph).

With these basics in mind we can define a Cayley graph. A Cayley graph is a digraph associated "with a group G and a set $S \subseteq G$ of generators" (Bretto et al. 2007: 55). S is some set of generators, or elements which *generate* the group. If G is a group and S is a subset of elements, then S generates G if every element of G can be expressed as a product of elements from S and their inverses. Returning to the abelian group $\mathbb{Z}$ or the integers with the addition operation for the moment, we can express every element as a sum of ones and their inverses: -3=-1-1-1 and 0 is the sum of 0 ones. So from this we can say that the group of integers is generated by the element 1.[24]

Lastly, the central notion of a symmetry is also representable in graph-theoretic terms. For instance, a dihedral is the symmetry group of an n-sided regular polygon for $n > 1$. Dihedral groups are non-abelian permutation groups for $n > 2$. And a permutation group is a finite group G whose elements are permutations of a given set with the group operation of composition of permutations in G. We have already seen the importance of the notion of permutation invariance in fundamental physics. Another way of talking about symmetries is with the idea of automorphisms, as Weyl himself states:

> [W]hat has indeed become a guiding principle in modern mathematics is this lesson: Whenever you have to deal with a structure-endowed entity Σ try to determine its group of automorphisms, the group of those element-wise transformations which leave all structural relations undisturbed. (1952: 144)

Σ is the symmetric group of all permutations of $[n] = \{1, 2, \ldots, n\}$. And this structure is quite easily presented in Cayley graphs. Furthermore, Cayley's Theorem shows that the abstract notion of a group and the notion of a group of permutations are basically equivalent. Meier (2008) further proves what he calls 'Cayley's Better Theorem' where "every finitely generated group can be

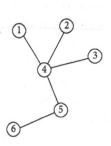

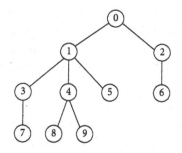

Fig. 7.1 Unrooted.   Fig. 7.2 Rooted.

faithfully represented as a symmetry group of a connected, directed, locally finite graph" (19).[25] The details are obviously beyond the present scope but they show an important correspondence between groups and graphs in terms of symmetries.

As we have seen, generative linguists use a particular kind of graph called a *tree* in their work. A tree is a connected graph with no cycles. This means that (1) there is a path from any vertex or node to any another [connected] and (2) that there are no paths that start and stop at the same node, but contain no other repeated vertices [cycle]. They tend to also be labelled with the names of phrases and terminals like NP, VP, N, V etc. Figure 7.1 is a labelled tree (I've just used numerals for the labels). Note most, not all, linguistic trees are also *rooted* as in Figure 7.2.[26]

Stabler and Keenan (2003) prove that isomorphisms of symmetry groups and invariance can play a significant role in the measurement of structural similarity within and across languages.[27] They start by defining grammars as tuples $G = (\Sigma, F, Types, Lex, \mathcal{F})$ where Σ is a non-empty alphabet, F a sequence of features, *Lex* is a subset of a complex Cartesian product called chains C composed of the alphabet, types, and features. Expressions are identified as C^+ and lastly $\mathcal{F}$ is the set of generating functions defined in terms of the Minimalist operations of *merge* and *move*. A language is defined as the 'closure(*Lex*, $\mathcal{F}$)'. From this they define both structure and symmetry in terms of syntactic automorphisms. I reproduce their definition 2 and theorem 2 below.

Definition 1: For any grammar G and any bijection h: $L(G) \rightarrow L(G)$, h is a syntactic automorphism (or symmetry) for $(L(G), \mathcal{F}$ iff for every $F \in \mathcal{F}$, $h(F) = F$.

Theorem 2: In any grammar G with generating functions $f^i, g^i \in \mathcal{F}$, if grammar G' is the result of adding $f^i \circ_k g^j$ to $\mathcal{F}$, these two grammars define exactly the same structure, in the sense that they define the same language, have the same symmetries, and have the same invariants.

They go on to use this technology to define structural similarity between categories and grammatical operations within and across different language structures. In fact, they even define the symmetry group in terms of a grammar. This is not surprising given the intimate relationship between grammars and graphs. But it is interesting that they choose to recreate group-theoretic structure for their structural analysis of language.

In fact, following notions of symmetry and invariance in physics, Johnson (2015) sets the precedent for the adoption of invariance considerations in the philosophy of linguistics, albeit for different purposes. He starts by modifying Chomsky on the notion of 'notational variants' or the idea that "two theories (formal grammars, etc.) are notational variants iff they are empirically equivalent" (Johnson 2015: 163) or following Chomsky (1972) do not differ in empirical consequences. He then presents a compelling case for applying a measure-theoretic analysis to generative linguistics. But before doing so he makes a few interesting points which verge on a structural realist view without endorsing (or mentioning) the possibility.

> Collectively, the notational variants of a theory determine the empirically 'real' or 'meaningful' structure of any one of the theories taken individually. This meaningful structure is often not identifiable without recourse to notational variants (i.e. symmetries). (Johnson 2015: 164)

He goes on to claim that notational variants can shed light on which parts of theories are of empirical consequence and which parts are mere artifactual structure. For instance, consider the difference between two ways of representing temperature, Celsius and Fahrenheit respectively. The 'real' empirical content or structure of temperature is determined by their convergence or intertranslatability. Anything *sui generis* about either system of representation is merely artifactual.

For a more controversial case involving linguistics, consider the discrete infinity postulate of generative grammars. If certain model-theoretic treatments of syntax do not entail cardinality properties (or are 'cardinality neutral' according to Pullum (2013)), then discrete infinity is an artifact of the formalisms used not a real feature of linguistic structure (see my (2019a) for a related argument). Johnson identifies the 'invariance principle' which roughly states that what is interesting empirically about a given formal grammar is not what it says but rather what it agrees with every other grammar on. This principle might be useful for providing an answer to the problem in Quine (1972) related to the psychological plausibility of multiple equivalent grammars (as we shall see), which is one target of Johnson's account, but in its strong form it also militates against a notion of scientific progress across generations of formal grammars. Thus, I would argue that certain so-called 'artifactual' or non-invariant structure can actually shed light on the differences and potential progress of later formalisms. In fact, the dispensability condition on real patterns advocated in Chapter 4 essentially requires non-redundant structure.

For instance, as reported by Bueno and Colyvan (2011: 364), multiple revisions, in terms of physical interpretations, of the same mathematical formalism in classical mechanics led to the discovery of the positron. Dirac initially thought negative energy solutions was merely features of the mathematical model and not physically realised but later, after finding physical interpretations of these solutions, it caused him to revise his entire theory and predict the existence of a novel particle. In general, the mathematical structures applied scientists use are much richer than the physical structures being modelled (and sometimes *vice versa*) and this can lead to predictions based on logical extensions of the mathematics or merely interpreting 'unused' mathematical structure.

Perhaps this is just to say that invariance is not the only means of identifying structural features relevant to understanding theory change. Nevertheless, it is a useful concept for identifying those parts of linguistic theory that have remained constant and those parts that have changed in a commensurable manner. In the last section, I will argue that it might additionally be useful as a tool for cross-formalism comparison.

7.4 The Problem of Multiple Grammars

Before concluding this chapter, I want to discuss the problem of multiple grammars initially posed against mentalism in linguistics by Quine (1972). A version of the issue can be reconstructed for the structural realist position I am advocating here. I offer a possible solution, in terms of 'multiple models idealisation' which asks us to think of structures in terms of scientific models, in the section after this one.

Quine's famous problem of multiple equivalent grammar formalisms goes something like this: take two weakly equivalent grammars, phrase-structure grammar and categorial grammar, for instance. These two grammars generate the same sets of sentences or in Quine's terms are "behaviourally equivalent". The problem is that since they are both empirically adequate or generate the same sentences, there is no way of deciding which grammar is the correct description of the target (in Quine's critique, the target would be mental states of language users). The same logic can be applied to the current analysis. If multiple grammar formalisms pick out distinct structures but have the same output or are weakly equivalent, how do we know which is real (in whichever ontological sense we prefer)?

This is also not a merely theoretical worry. Recently, there have been a flurry of formal proofs of weak or expressive equivalence of various syntactic formalisms such as tree adjoining grammar (Joshi), generalized phrase structure (Pollard) and categorial grammar (Steedman). Furthermore, generative syntax can also be shown to be equivalent to these formalisms.[28]

Of course, Quine's argument is a subspecies of the larger issue in that of 'underdetermination of theory by evidence' in the broader philosophy of science. When two *empirically equivalent* scientific theories converge on the data, then this poses a problem for the scientific realist as to which theory is accurate of the world. The issue is iterated at various levels. As van Fraasen puts it:

> The phenomena underdetermine the theory [...] The theory in turn under-determines the interpretation. Each scientific theory, caught in the amber at one definite historical stage of development and formalization, admits many different tenable interpretations. What is the world depicted by science? (1991: 491)

One way of escaping this issue is by appeal to shared underlying structure. Another would be specifying structural overlap in cases in which the structures of the respective theories cannot be identified (see French 2014, chapter 2 for a number of options for locating a unifying framework). Generative linguists, following Chomsky (1965), have often adopted a realist strategy of finding a 'breaking factor'. In other words, *descriptive adequacy* (i.e., coverage of native speaker grammaticality judgements) is replaced with a criterion of *explanatory adequacy* which includes language acquisition data. This strategy just passes the buck though and pushes the problem down the line. Thus, additional criteria are sought to break later deadlocks, such as the Minimalist condition 'beyond explanatory adequacy' or 'natural adequacy' (Chomsky 1995, Hinzen 2012) which adds further evolutionary factors. We'll consider more structural

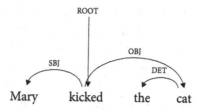

Fig. 7.3 Dependency graph for the
sentence *Mary kicked the cat*.

resolutions to the problem here as I think the standard strategy suffers from
similar issues to that of simplicity and explanatory power 'breaking factors'
in the general philosophy of science. Not to mention what counts as an
evolutionary deciding factor can depend heavily on one's chosen framework.

Let us consider two frameworks, both alike in dignity, which eschew some
of the core tenets of generative grammars. Firstly, dependency grammar is
a family of grammatical formalisms which have received renewed interest
in linguistics as well as in applied linguistic settings such natural language
processing (NLP), psycholinguistics, and neuroscience. It is based on a notion
of *dependencies* which are binary, asymmetric governance relations which hold
between words. If word A dominates or governs word B, then word B depends
on word A. In standard terminology, word A is called a *head* and word B a
dependent. These relations are labelled in dependency trees, which are rooted,
directed acyclic graphs. One node acts as a root, from which arrows (relations)
emanate to other nodes in the structure. Importantly, nodes are in a one-
to-one mapping with words. So dependency grammar has just one layer of
representation or 'flat structure' as opposed to the hierarchical structures of
phrase structure grammars. Below is a standard dependency graph.

Dependency grammars are easier to parse than phrase-structure grammars,
as they tend to reflect argument structure and semantics more clearly. This is
in part due to their use of functional labels, as well as the fact that they
do not contain phrases, recursion explicitly, or involve movement. But they
are equivalent formally to phrase-structure grammars or rather context-free
grammars (proven by Hays 1964) thus they have many of the expressive
capabilities of more complex formal languages.

Of course, the consideration of dependency grammar which incorporates
flat structure and no overt recursion or movement opens the door to alternative
frameworks such as those which derive from the other side of logic, namely
model-theoretic syntax. I gave a brief introduction to this concept in the
introduction and added some details in Chapters 4 and 6. Under those

conceptions of syntax, the field is more related to formal languages on the model-theoretic side of logic as opposed to the proof theory which inspired much of early generative linguistics (Pullum 2011). Pullum (2013) takes this to be at the heart of what he calls "syntactic metatheory" or the metascientific study of frameworks for natural language syntax.

> The question I regard as most central, in that it arises at a level more abstract than any comparison between rival theories, concerns the choice between two ways of conceptualizing grammars. One has its roots in the mathematiciza- tion of logical proof as string manipulation in the early twentieth century, and the other springs from a somewhat later development in logic, namely model theory. (492)

Model-theoretic syntax replaces the centrality of that proof-theoretic notion of syntax with a more logical syntax drawing from first-order logic and model theory.[29] As Morrill (2011: 64) states "when the formal systems of syntax resemble deductive systems, we may speak of logical syntax in a strong sense". Grammar formalisms such as Generalized Phrase Structure Grammar (Gazdar et al. 1985) and Head-Driven Phrase Structure Grammar (Pollard and Sag 1994) are prominent examples of this approach. They are contrasted with the derivational approach of generative grammar. Simplistically, instead of a gram- mar generating the well-formed formulas of a language, a grammar models the rules of that language via constraints. "An MTS [model-theoretic syntax] grammar does not recursively define a set of expressions; it merely states necessary conditions on the syntactic structures of individual expressions" (Pullum and Scholz, 2001: 19). In this sense, a sentence or expression is well- formed iff it is a model of the grammar (defined in terms of constraints which act as the axioms of the formalism). To be a model of the grammar is to be an expression which satisfies the grammar or meets the constraints.

Interestingly, from a formal point of view, model-theoretic techniques can characterise generative grammars along the Chomsky Hierarchy as we saw in the previous section (and in Chapter 4). Chomsky (2000b) even considers derivational and constraint-based approaches to syntax to be notational variants of one another as we saw above. Again, as we saw in Chapter 4, a clue as to how to approach the question of structure might be brought out by the distinction between *weak* and *strong* generative capacity in formal language theory. There we suggested that the latter captures the essence of linguistically real patternhood. Recall that weak generative capacity can be characterised in terms of the class of formal languages generated by a specific grammar in the Chomsky Hierarchy. Strong generative capacity, on the other hand, not only assigns a class of formal languages to a given grammar, but also a structural description. Thus, in terms of strong generative capacity, a dependency grammar is associated with a rooted directed acyclic graph as stated above, but phrase-structure and HPSG trees are not.

One could adopt the invariance approach suggested by Johnson above. Dependency grammars, model-theoretic grammars, and generative grammars might share some invariant structure, i.e. the 'true' empirical content of the theories.[30] But as mentioned before, this approach can miss genuine theoretical and even relevant structural differences between formalisms, i.e. linguistically real patterns. Thus, I think the invariance approach can be supplemented with a multiple models approach (we'll see another case for this in Chapter 9 when we discuss integration into cognitive science).

7.4.1 Multiple Models Idealisation

The possibility of multiple models firstly involves not thinking of grammars as theories at all, of either mental competence or otherwise. I have argued for this view elsewhere (see my 2016, 2019b) and it was implicit in much of the discussion towards the end of Chapter 4. The literature on scientific modelling is vast (see Giere 1988, Morgan and Morrison 1999, Godfrey-Smith 2006, Weisberg 2013). This alternative has it that grammars are best seen as scientific models or *indirect* representations of the target system in which a form of 'surrogate reasoning' is employed (Suárez 2004).

Multiple and even incompatible models would then receive interpretation under the banner of what is called 'multiple models idealisation' (Weisberg 2007) in the literature. This practice involves constructing many connected but potentially incompatible models each of which focuses on one or more aspects of the target system. This strategy differs from other kinds of idealisation "in not expecting a single best model to be generated" (Weisberg 2007: 646). Naturally, strong structure preservation or one-to-one correspondences are not appropriate within this practice. Since scientific theories can have diverse goals such as accuracy, simplicity, predictive power, etc., and the construction of one model to fit all of these criteria necessarily involves 'trade-offs', this approach offers the theorist a way of meeting all of these objectives separately.

If a theorist wants to achieve high degrees of generality, accuracy, precision, and simplicity, she will need to construct multiple models. (Weisberg 2007: 647)

This modelling practice is common in climatology, ecology, biology and population studies. It also fits well with the systems biology approach suggested in Chapter 5. If we consider the various models used in the service of linguistic theory and aimed at natural language, this might offer us a means of unifying different formalisms in their diversity. Dependency grammars map different syntactic relations than do phrase-structure. They track semantic relations better and are more parsable. Phrase-structure models track constituency and transformations such as passive to active or WH-movement better. Model-theoretic approaches can better capture notions of ungrammaticality (only derivationally determined by generative grammars) and so on. In other words, each formalism picks out indispensable real patterns. There is even growing evidence to suggest that we might process natural language syntax in terms of separate syntactic modules that keep track of different kinds of structures (Lopopolo et al. 2020, Nefdt and Baggio 2023). Interpreting grammars as models which indirectly pick out structures can provide us with the additional tools needed for inter-theoretical comparisons in terms of structure.

It should be mentioned that accepting multiple models does not amount to pluralism without qualification. Hasselman et al. (2015) specifically advocate structural realism as a means of unifying pluralist methodologies and ontologies in cognitive science. They claim that structural realism avoids ontological reductionism and offers a means of unifying disparate elements of the cognitive scientific landscape. Their focus is mostly on the epistemic variant of the theory though. In Chapter 9, we will return to the issue of cognitive science and ontic structural realism. The authors further claim, among other things, that

> [T]he SR stance provides us with adequate and sound tools for unifying, discarding or even building theories in the process of theoretical advancement of the field. The 'discovery' of a structural connection between a set of phenomena or between existing theoretical constructs, driven either by converging empirical evidence or by conjectured theoretical relationships, is what scientific progress is all about. (Hasselman et al. 2015: 11)

Similarly, the multiple models idealisation suggested here does not preclude the possibility of a unified ontology at the metatheoretical level. Structural realism offers one means of achieving this goal as I have argued in this section and above.

Furthermore, French (2011) makes a strong case for the adoption of the models as structures approach as a conduit towards structural realism in biology. In response to the concern that biology does not involve the kinds of equations and laws of physics, thereby making structural realism less applicable, French states "in the apparent absence of the kinds of laws that structuralists in physics can point to, biological models can play the same role in offering us access to the biological structures of the world" (2011: 167). Again, in response to the concern that fields like biology (and by extension linguistics) involve a diversity of models (Odenbaugh 2008) or a 'patchwork' picture of loosely stitched together combinations of various tools, French follows Hacking (1983) in pointing out that physics too can be seen in this light. In biological explanation, weakly interacting variables are sometimes unavoidable and so too multiple models. But accepting some disunity is not tantamount to giving up on structuralism in so far as we could "accept that we cannot arrive at a completely unified structuralist representation of the world but that each of the patches or facades represents some piece of the underlying structure" (French 2011: 169). In addition, such acceptance does not mean that the structure of world is similarly fragmented.

7.4.2 Multiple Grammars and Multiple Reductions

Lastly, there is an obvious parallel here between the problem of multiple grammars as I have called it and one of the primary motivations for structuralism in the philosophy of mathematics, namely 'the problem of multiple reductions'. This issue was initially presented in Benacerraf (1965) with relation to arithmetic. The purported relationship between systems of objects and structures is often said to be captured by the process of 'Dedekind abstraction' (as we saw in Chapter 4.2). The idea is that a *structure* is an abstract form of a system which homes in on only the structural (relational) properties of the objects of the system and nothing else.

It is important to remind ourselves that this notion of abstraction has ontological import. Unlike in modelling literature, where abstraction is usually understood as a tool used mostly for tractability or simplicity, mathematical structures (the target of mathematics as a science) *are* purely structural in nature, i.e. they only have structural properties, on this view. For example, mathematical structuralism, especially the *ante rem* variant (as we've seen in Chapter 4), states that both the Zermelo numerals and the von Neumann ordinals pick out the same structure, i.e., the natural number structure. Thus, the corresponding claim would be that multiple grammar formalisms pick out same linguistic structures or LRPs. But we know that they do not, at least in terms of strong generative capacity.

Furthermore, the question of the identity of these mathematical systems is a tricky one. Identifying them would be tantamount to ignoring set theoretical differences and making a clear distinction between essential and non-essential features of mathematics. Shapiro (1997) opts for identifying systems 'up to isomorphism' (but later Shapiro (2006) vacillates on this point). Resnik (1997) avoids talk of identity altogether. Nevertheless, there might be a certain revenge of Benacerraf's problem for structuralism similar to the case of grammars we are discussing. Following Hellman (2001), Gasser (2015: 17) states:

> One might object to this response by pressing the Benacerrafian thought one step further. For given any *structure* which preserves the abstract form of a system of objects related to each other by a successor-like relation, other structures will exist which can do the job equally well.

And we are back to the Quinian situation, this time for mathematics itself. Gasser (2015) motivates for a non-ontologically committing version of structuralism in the wake of this possibility. However, the modelling analogy is available to the philosopher of mathematics similarly. If the Zermelo numerals and von Neumann ordinals are viewed at the level of models, the idea of *indirect* tracking of the natural number structure potentially avoids the Benacerraf-type worries.[31] This, of course, is a task for another day (and perhaps another philosopher). In chapter 9.6, we'll uncover another face of this structural conundrum in cognitive neuroscience. For now, we can appreciate that there are many options for dealing with the challenges we set out at the beginning of this chapter and that structural realism, buttressed with a scientific modelling perspective, is a viable foundation for the philosophy of linguistics.

8

Language at the Interface

So far the trajectory of the book has followed the focus of other works in the philosophy of linguistics by aiming its philosophical gaze largely at syntax. By contrast, the philosophy of language has traditionally reserved a much more starring role for semantics. In recent years, metasemantics has emerged as the prominent pursuit of metatheoretical inquiries into the nature of meaning and its scientific study thereby stretching the traditional province of the philosophy of linguistics. Of course, the philosophy of phonology, pragmatics, and morphology are also natural extensions of the purview of the field, but for various reasons they have been underexplored (see Nefdt 2023).

My focus on syntax thus far was not without precedent. Syntax or the study of linguistic form took center stage due to its advanced stage of formalisation and formalisability. This led many, including Chomsky, to consider syntax the most worthy candidate for naturalisation or scientific study. The situation was compounded by advances in proof theory and the rise of computationalism in the cognitive sciences. The result was that semantics (and other subdisciplines) were relegated to the realm of unscientific noise at the worst times. Sampson (2001) even goes as far as to suggest that semantics is akin to fashion or art in its unsystematic nature. But as we shall see *one linguist's noise is another's pattern.*

In fact, one way of perceiving the long fought 'linguistic wars' between generative syntacticians and the then emerging field of generative semantics is that semanticists dared to find patterns in the linguistic morass (Harris, 2021, see). Of course, the real issue was about whether or not those patterns belonged to the linguistic system or went beyond it. In the following chapter, I will assume that semantics, phonology, pragmatics all exhibit structure in different ways. In fact, they all contribute to the linguistically real patterns identified in Chapter 4. In that chapter, and its chief measurement tool of formal language theory, structure came for free to a certain extent. Hence the focus on syntax as the field of linguistic theory that best exemplified both the real pattern analysis and the structural realist paradigm it invited.

In the following chapter, I expand the focus to semantics as well as the interfaces between the various subdisiplines and the systems they represent.

Language, Science, and Structure. Ryan M. Nefdt, Oxford University Press.
© Oxford University Press 2023.
DOI: 10.1093/oso/9780197653098.003.0008

Specifically I plan to exploit a feature of informational structural realism (Floridi, 2008a,b, 2011) to present a new take on the interfaces between linguistic domains and the patterns they exhibit in terms of 'the method of levels of abstraction' (Floridi and Sanders, 2004). The idea of levels of analysis or interpretation, and the method upon which it is based, should be familiar from Dennett's hierarchy of stances and Ladyman and Ross' scale relativity of ontology, both of which we've seen already.[1] The latter thesis was pitched at the ontological level, however. Here, I plan to focus on the epistemic version of the idea while commenting only briefly on the ontology of semantics. The central insight of this chapter is thus:

CENTRAL INSIGHT VII: NATURAL LANGUAGE IS A COMPLEX SYSTEM CONTAINING LRPs ANALYSABLE AT DIFFERENT 'LEVELS OF ABSTRACTION', EACH CONNECTED BY A NESTED 'GRADIENT OF ABSTRACTION'.

Importantly, CI VII is an epistemic claim about how the relationship between different linguistic systems are modelled, *via* different subdisciplines, not a claim about their ontology. This insight might seem heterogeneous at first glance but our approach will be systematic in what follows. Specifically, what is meant by 'levels of abstraction' will be divorced from intuitive understandings and given a technical interpretation. In addition, 'gradient of abstraction' is also technical term to be explained in the next section. But the basic idea is that phonology, syntax, semantics, and pragmatics are all different levels of abstraction (again, to be defined below) targeted at the complex system of language. Therefore, their interfaces are best modelled as incremental information growth of connected but independent patterns resulting in more and more fine-grained analysis and measurement at higher levels.

I also owe the reader some detail on what I take to be a 'complex system'. I've done this to a certain extent by analogy with biological systems which are complex systems (in Chapter 5). Here, I'll say a brief word on the more high-level definition of a complex system based on recent work by Ladyman and Wiesner (2020), before continuing with the topic on the chapter.

8.1 A Note on Complex Systems

Most of us have an intuitive concept of what a complex system is. Some organic or inorganic assemblage of independent and interacting component parts which work together somehow to perform tasks or survive (usually both).

The problem with leaving it at this vague statement is that almost anything can be considered a complex system if this were our guiding insight. In fact, one of the reasons the science of complex systems has taken so long to emerge is precisely because many scientists and philosophers believe that the target phenomenon is too amorphous for precise study (notwithstanding, the reductionism that held sway for a while). Chomsky certainly had this worry about extending the science of linguistics beyond the uniformity and elegance of syntactic structures (for the language faculty).[2] The result, in linguistics and science more generally, is that idealisation silos are created to study individual aspects of a complex target with more targeted methodology. This methodology can have a Medusian effect on the study of a particular object or structure. If you look too closely with instruments of maximum precision, you can quickly become convinced that the object of inquiry is exactly the thing for which the tool has been designed by some cosmic coincidence and lose sight of the larger picture. Jackendoff (2002) cautioned that the initial Chomskyan idealisation of the competence-performance distinction has hardened over time so much so that the complex system of language was out of sight. More specifically, the competence model became impossible to reconcile with linguistic performance within traditional generative approaches, at least according to his critique.

In more philosophical parlance, laser-pointed focused attention can lose sight of 'emergent properties'. Clark (1996), for example, views language as a coordinative dance only truly emerging within the setting of multi-person performance of communicative acts. Dynamic syntax similarly takes dialogue data very seriously in its modelling of syntax (and inextricably semantics). So what does it mean to be a complex system, and how do we study such things?

A full discussion of these fascinating questions would take an entirely different book (see, the beginnings of this kind of view for language sciences in Nefdt (forthcoming)). Luckily, there has been a recent such offering from Ladyman and Wiesner (2020). There they provide an overview of the field of complexity science, its history, prominent examples, and methodological underpinnings. Curiously, on the list of the complex systems they describe, such as the human brain, bee colonies, the universe and the climate system, they leave natural language out of the picture. Perhaps this book can be seen as a defence of why it should be included in the list.

Specifically, there are a number of properties or features of complex systems that I think natural language fits and a number of aspects of complexity science that I think linguistics (broadly construed) exemplifies. With regard to the latter, Ladyman and Wiesner (2020: 9) identify nine truisms about complexity science, five through to eight are especially relevant to us.

5. Complex systems are often modelled as networks or information processing systems.
6. There are various kinds of invariance and forms of universal behaviour in complex systems.
7. Complexity science is computational and probabilistic.
8. Complexity science involves multiple disciplines.

In Chapter 5 I argued explicitly for the first disjunction in (5) and in this chapter I will show some evidence for the second at the interfaces. Chapter 7 made a case for (6) and the role of invariance in linguistics. Chapter 4 expounded on the computational elements of natural language and the present chapter will add the probabilistic aspects (particularly with relation to metasemantics below). The book as a whole (and especially Chapter 9) advocates for the multidisciplinarity of the study of language.[3]

Sadly, its beyond the present scope to motivate more strenuously for the view that linguistics is or should be a complexity science (although Chapter 5 does go some way in doing that). Nevertheless, one of the core features of a complex system identified by Ladyman and Wiesner (Anderson, 1972, following) is that they often exhibit 'nested structure and modularity'.

> There is nested structure in the physical world from the subatomic through the atomic to the chemical, and ultimately to planets, stars, galaxies, clusters and superclusters. The Earth and its oceans and atmosphere exhibit a very rich nested structure, as does the solar system. In such complex systems there is structure, clustering and feedback at multiple scales. (Ladyman and Wiesner, 2020, 81)

Of course, structure has played a central role in the dialectic of the book so far (and it will continue to do so until the end). The further concept of 'nested structure', however, needs development. The same applies to the related property of modularity or the "functional division of labour, or specialisation of function among parts" (Ladyman and Wiesner, 2020, 107). This latter property is well-attested in linguistics since generative grammar (see the Introduction) and a number of other frameworks discussed in this chapter. Thus, the present chapter aims to explore an account of the nested structure and modularity of the linguistic interfaces.

8.2 Levels of Abstraction

Before defining the method of levels of abstraction, I want to briefly describe the background theory from which it springs (in the philosophy of science, at least). Informational structural realism (ISR) is a view largely informed by computer science and the field of 'formal methods' therein. It has a number of aims, the first of which is to reconcile epistemic structural realism with its ontic sibling. The second aims to improve upon the appeal of structural realism more generally by proposing an account of the nature of the 'relata' of structures, i.e., rescuing the concept of structural objects. This is something Floridi (2008a) claims is lacking in the literature. What's left is an account which dances on the line between instrumentalist or 'internal realism' and 'metaphysical realism' (what Floridi calls 'liminal realism'). The resulting worldview evoked is "a version of OSR [ontic structural realism] supporting the ontological commitment to a view of the world as the totality of informational objects dynamically interacting with each other" Floridi (2011, 340). We won't be focused on the general philosophy of informational theory directly in what follows (for more see, Floridi, 2011). What we will take from Floridi's account is a framework for relating the various ways of connecting and relating systems in terms of their different abstraction statuses.

The central tool for defining ontological commitment in ISR is the 'method of levels of abstraction' which is a modelling technique developed in computer science to analyse informational systems by means of discrete mathematics. It's specifically prevalent in Object Oriented Programming (OOP).[4] As Greco et al. (2005) state:

> [t]he process of making explicit the Level of Abstraction (LoA) at which a system is considered is called *Method of Abstraction*. This method pertains to the analysis of discrete systems, by which we mean those systems whose evolution is described by explicit transition rules. It applies both to conceptual and to physical systems. Its pivotal element is the concept of LoA. (627)

The method itself consists of three important concepts, (1) levels of abstractions (LoAs), (2) system behaviours, and (3) gradients of abstractions (GoAs). The first step, or (1), involves identifying the 'observables' or 'interpreted typed variables' of the system. In a theory of syntax, these could be the set NP, VP, PP that corresponds to the distributional patterns of a given language. Different languages might have different categories but as long as there is a finite, non-empty set of these typed variables we are at a level of abstraction. In the P&P model, Chomsky called the typed variables and objects constructed from them 'primes'. The observables would then be 'phrase markers' as in the categories above usually represented by trees or labelled brackets (as I have done) (see Chomsky 2015: 30). There is, however, some flexibility on the members of the sets.

> Being an abstraction, an observable is not necessarily meant to result from quantitative measurement or even perception. It does not have to be a physical magnitude either. Although the "feature of the system under consideration" might be empirical and physically perceivable, an observable might alternatively be a feature of an artefact or of a conceptual model, constructed entirely for the purpose of analysis. (Floridi, 2008a, 226)

Importantly, given a specific interpretation and context, certain combinations of observables become unacceptable. In a given sentence of English, all the values can't be PP as this would be an unacceptable string. Floridi (2008a,b) example involves colours of traffic lights. Four traffic lights modelled by four observables cannot all be green together (unless some defect is present). The mechanism for constraining the possible combinations of values is classed under 'system behaviours' which describes only the permitted combinations. In terms of (2), more formally "[a] *behaviour* of a system, at a given LoA [level of abstraction], is defined to consist of a predicate whose free variables are observables at the LoA" (Floridi, 2008a, 227). Substituting values for variables results in a true predicate of 'system behaviours'. Specifically in our case, an LoA can have as input a particular *system* like syntax, comprising *data*, i.e. distributional patterns. It's output is a *model* which contains syntactic *information*.

(1) and (2) combine to form 'moderated LoAs'. So a finite non-empty set of observables coupled with constraints on the possible combinations of values of variables is a moderated LoA. Here's another linguistic example. Verbs take arguments. In simple settings, the most common kinds of arguments are *agents* and *patients* (which are both NPs). In the sentence *Nomusa wrote the letter*, *Nomusa* is the agent (performing the action) and *the letter* is the patient (being acted upon). Passive constructions allow for agents to be optional in most languages. Intransitive verbs lack patients. But all verbs in non-pro drop languages like English require arguments (so-called 'avalency' is not permitted), except in rare situations like expletives. Therefore, for such languages, if the variable a represents arguments (of verbs), the predicate designating the value of the variable cannot be zero (or negative) in number. If a value falls within the $1 < a < n$ interval, then it is a system of behaviour.

The idea is that the behaviour of the system moderates the LoA. So there are constraints at every level. We start with typed variables which are variables constrained in terms of their type, then we move to observables with are further interpreted typed variables and LoAs which are collections of these observables constrained by a delimited space of possible values or the system behaviour. Finally, the moderated LoAs are LoAs predicated on this behaviour.[5]

Take the case of temperature measurements. Two scientists, one partial to Fahrenheit calculations and the other to Celsius, might use the same typed variable t but the observables of their theories would differ since they represent values differently, i.e., they interpret the variables differently. The range of acceptable values would also differ, since 0° Celsius is equivalent to 32° Fahrenheit, so the system behaviour also differs. Of course, we can relate the different systems *via* translations as their correspondence is systematic. More on translation below.

With the concept of moderated LoAs in hand, we can go on to define the most essential tool within the method under discussion, namely (3) 'gradients of abstraction' or GoAs. The function of GoAs is to "facilitate discussion of discrete systems over a range of LoAs" (Floridi, 2008a, 227). The LoAs represent the scope of a model and the GoA provides a way of altering the LoA in order to make observations at different levels of abstraction. The GoA is in a sense another constraint on LoAs: that the observations at one level explicitly relate to those at all the others. In other words, observations at the level of syntax have to be related to those at the level of semantics. But we are getting ahead of ourselves. Floridi (2008b) uses a more vinous example.

> For example, in evaluating wine one might be interested in the GoA consisting of the "tasting" and "purchasing" LoAs, whilst in managing a cellar one might be interested in the GoA consisting of the "cellaring" LoA together with a sequence of annual results of observation using the "tasting" LoA. The reader acquainted with Dennett's idea of "stances" may compare them to a GoA. (311)

The observables which constitute "tasting" LoAs might be things like "acidity", "clarity", "colour" and so on while those at the level of "purchasing" might be "vintage", "region" and so on where the South African origin of the wine might make more of an impact on buying patterns. Storing or "cellaring" again involves other observables. There might be some overlap in typed variables but their interpretation would be slightly different (e.g., "vintage" might be relevant at different levels). The mandate, however, is that observations at each LoA be "explicitly related to those at the others" (Floridi, 2008b, 311). In order to do so, we need a family of relations and two conditions to begin with. I reproduce the definition (from Floridi, 2008a,b, and elsewhere) below:

A gradient of abstractions, *GoA*, is defined to consist of a finite set $\{L_i \mid 0 \le i < n\}$ of moderated LoAs L_i and a family of relations $R_{i,j} \subseteq L_i \times L_j$, for $0 \le i \neq j < n$, relating the observables of each pair L_i and L_j of distinct LoAs in such a way that:

1. the relationships are inverse: for $i \neq j$, $R_{i,j}$ is the reverse of $R_{j,i}$
2. the behaviour p_j at L_j is at least as strong as the translated behaviour $P_{R_{i,j}}(p_i)$

A 'translation' from one LoA to another is a relation that takes any predicate p which moderates one and finds its image which satisfies the predicate in the other LoA. The result is that condition (1) ensures that the behaviour moderating LoAs at lower levels is consistent with behaviour at the higher ones. Moreover, "[t]he consistency conditions imposed by the relations $R_{i,j}$ are in general quite weak" (Floridi and Sanders, 2004, 16). Given this weakness, we can specify the kinds of GoAs in order to more accurately model the relationships between the LoAs we deem appropriate for a target phenomenon or set of phenomena. This is the task of the next section.[6] It is also the point at which the gradient of abstraction dovetails with theses like the more ontologically driven scale relativity thesis of Ladyman and Ross or Dennett's stances, since it allows us to consider the same system from different perspectives.

To conclude, the method of abstraction makes explicit and formal a number of views on various topics in metaphysics, the philosophy of mind, and other fields. By drawing on practices in computer science and ISR it also offers us new ways of assessing discrete systems like language (or at least how it's been predominantly modelled), as I hope to show below.

8.3 The Proposal

In this section, I plan to offer a way of appreciating the interfaces between different linguistic systems and the models they produce in terms of the method of abstraction discussed above. Greco et al. (2005) argues that LoAs are comparable to a different kind of interface, namely of computer systems.

The metaphor of interface in a computer system is helpful to illustrate what a LoA is. As is well known, most users seldom think about the fact that they employ a variety of interfaces between themselves and the real electro-magnetic and Boolean processes that carry out the required operations. An interface may be described as an intra-system, which transforms the outputs of system A into the inputs of system B and vice versa, producing a change in data types. (627)

There are similarities between this idea of an interface and the various kinds of linguistic architectures that have been proposed in generative grammar and other frameworks (see below). But the reason Greco et al. use the metaphor is that like interfaces, LoAs use a network of observables moderated by behaviours to define a place where "independent systems meet, act on or communicate with each other" (2005: 627). The last point is where things get interesting. Chapter 5 pushed an account of natural language as a complex system. In this section, I want to make more precise the relationship between the constituent parts of this system.

There are two different kinds of GoAs I think appropriate for linguistics. The first is one we've seen in Chapter 7 already in a less abstract guise and it is *inter-theoretical*. So I won't go into too much detail here. The second, and my focus, is *intra-theoretical* and I will use it to describe the interfaces between models of different linguistic systems as well as a comparison between opposing linguistic architectures.

The literature on GoAs specifies two salient types: (1) disjoint GoAs and (2) nested GoAs. Disjoint GoAs characterise systems with minimal to no overlapping parts or in the present parlance: no common observables. This picture captures nicely the situation with multiple grammar formalisms and the multiple models idealisation for which I advocated at the end of Chapter 7. There, I claimed that different grammar formalisms track complementary aspects of the target system. One formalism could track constituency while the other better represents argument structure.

Similarly, in semantics there are many distinct traditions. Mainstream model-theoretic or truth-conditional accounts, following Montague (1970) and Lewis (1970), pair set-theoretic objects with sentential level phenomena like anaphora, quantification, conditional constructions, etc. The central tool of this enterprise is functional application and the overarching framework assumes the principle of compositionality (Szabó, 2000).[7] Both of which are mappings from syntactic operations and structures to their semantic counterparts. This is on the discrete side of things. Distributional approaches challenge the centrality of compositionality and the growing popularity of vector-space models of meaning indicates the probabilistic and continuous aspects of natural language meaning are receiving more attention (see Erk, 2020). The model-theoretic approach seems to capture some correspondences with formal syntactic structure while the distributional approaches are skilled at spotting collocational and contextual semantic effects. A common adage found in the foundations of these latter research programmes is that "you shall know a word by the company it keeps" (Firth, 1957, 11).

A disjoint GoA would be appealing as a mechanism to capture the broad picture in semantics which involves LoAs at the level of truth-conditional approaches and distributional ones. While the various projects aimed at reconciling them would be instances of attempting to find more integrative GoAs (see Liang and Potts, 2015).

One such 'integrative' GoA is the *nested* variety, which I will suggest can better serve the characterisation of intra-linguistic interfaces. This GoA models incremental information growth. It comes with a formal condition that "the reverse of each relation $R_{i,i+1}$ is a surjective function from the observables of L_{i+1} to those of L_i" (Floridi, 2008a, 228). By stipulating this we get a kind of hierarchy in which the observations at higher levels map onto at most one observable from lower 'more abstract' levels but not *vice versa*. This captures a certain monotonicity of information growth in the overall system. Thus lower level observables can map onto many different higher level observables. For instance, if a syntactic observable (e.g. grammatical rule or category) is ambiguous or allows for multiple meanings, then it maps onto multiple observables at the semantic LoA. But the syntactic information is still contained in that latter system. Similarly, as we know from Grice onward, the same semantic representation or observable can be used in various contexts to convey distinct pragmatic contents or uses.

My proposal in essence is that we treat the linguistic interfaces between syntax, semantics, and pragmatics as a *nested* GoA in which each individual LoA is a model of linguistically real patterns of each respective system. The observables are drawn from the standard observables used in a given subdiscipline. For example, syntax uses NPs, VPs, structural relations like c-command and *Merge*, etc. The semantic LoA contains some of the same typed variables like NP, VP, etc., then adds structural relations corresponding to c-command and *Merge* like scope and functional application but since they receive different interpretations they constitute different observables. Of all the possible (possibly infinite) syntactic structures at the syntactic LoA, only a subset of these receive meaningful interpretation at the higher semantic

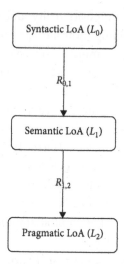

Fig. 8.1 Nested GoAs

LoA. Some syntactic structures (such as certain empty categories) won't fall within the behaviour of the semantic LoA at all (rendering the predicates of the moderated LoA false at that level). I'll present the idea in the diagram above, a nested GoA with three LoAs, before complicating it further.

This simple picture represents the flow of information within a particular theory of language. For instance, if the LoA at L_0 is generative syntax, the LoA at L_0 is standard truth-conditional semantics and the LoA at L_2 is a Gricean approach to pragmatic content. $R_{0,1}$ might be a homomorphism (often assumed in the compositionality literature) mapping syntactic rules to semantic ones. Each level contains its own observables and behaviour but there are relations that funnel a subset of that information to the next level. An additional feature of this setup is that it allows individual LoAs to contain more information than is strictly necessary for translation at other levels. In other words, there can be syntactic behaviour that receive no semantic overlay and there could be purely semantic and/or pragmatic observations not based on syntax. Syntax can be considered the most 'abstract' level of linguistic representation since it involves structural LRPs. But that structure is contained at L_1 and L_2 with the latter being the most detailed level of abstraction (and also the messiest).

If we zoom out to a bigger picture of linguistic theory at large, we can embed more disjoint GoAs within the overall nested structure. By doing this, we allow for multiple LoAs for syntax, semantics and pragmatics with multiple paths

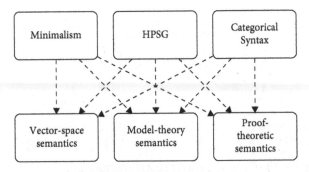

Fig. 8.2 Nested disjoint GoAs.

of potential connection. In a sense, this offers us a way of merging the inter and intra-theoretical into one framework. For instance, you could pair model-theoretic syntax with model-theoretic or proof-theoretic semantics favoured by inferentialists (or even a distributional approach). The LoAs at each level form a disjoint GoA but the overall GoA is nested.

In Nefdt (2020b), I presented a related picture (*sans* the method of levels of abstraction) of natural language semantics as an inferentially nested and iterated set of models connected to syntactic models. Although the accounts are compatible, what I am offering here goes beyond the previous one to specify how disjoint models might be embedded within a nested system. The current view also does not require semantics (or pragmatics) to be wholly parasitic on syntax. It highlights possible relations (indicated by dashed lines in Figure 8.2.) but also the independence of each model.

To close off the proposal I want to apply the above framework to three well-known linguistic architectures. The first two frameworks, Minimalism and the Parallel Architecture (PA) respectively, exemplify nested GoAs while the last one, Lexical Functional Grammar (LFG), shares some elements with the disjoint plus nested picture I have indicated above.

Minimalism (*circa* 1995–present) is the dominant iteration of the generative programme in linguistics. Practitioners often insist that it is not a theory but a 'program', a set of guiding principles, all based around the evolution-linked concepts of virtual conceptual necessity and simplicity. We've already seen various elements of minimalist accounts, their focus on the *Merge* operation as linguistically unique and central, their explanatory aims beyond native speaker judgements and acquisition data and the focus on evolutionary development. As was argued in Chapter 7, much of the structure of contemporary generative linguistic analysis is common or similar across various epochs of the older versions of the theory. In a similar vein, the architectural schemata have

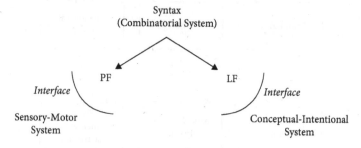

Fig. 8.3 The General Generative Architecture.

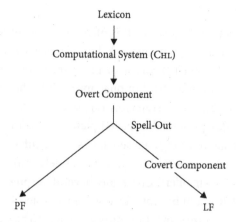

Fig. 8.4 Minimalist Architecture.

remained relatively constant. Syntax has always been the central computational fulcrum of the theory, while phonology and semantics (under various guises like 'Logical Form') have remained almost epiphenomenal. The Minimalist architecture of the language faculty, as with earlier versions, looks something like the Figure 8.4 above with more focus on the lexicon than previous accounts.

In the general picture, syntax is the most abstract LoA, with PF (Phonological Form) and LF (Logical Form) emanating from this system. It is nested because PF and LF contain syntactic information translated into different data types for consumption by their respective cognitive systems. More specifically, in Minimalist theories, there are two kinds of interfaces, the *internal* and the *external*. The lexicon and the syntactic or computational system constitute the former. This is where *Merge* takes elements of the lexicon and generates derivations. These derivations are, then, in turn read like instructions for the external systems, SM and CI, in Figure 8.3. PF and LF are external 'interface levels' with the result that a given linguistic expression is "nothing other than a formal object that satisfies the interface conditions in the optimal way" (Chomsky, 1995, 171). In Figure 8.4, there is a further distinction between *overt* and *convert* syntax. As Müller (2018: 128) explains "[o]vert syntax stands for syntactic operations that usually have a visible effect. After overt syntax the syntactic object is sent off to the interfaces and some transformations may take place after this Spell-Out point". However, these transformations are hidden from PF so they form part of the covert component.[8] In other words, formal abstract data structures are created by the shared LoA of the syntactic combinatorial system and the Lexicon. Once these structures are in place, they undergo transformation into more complex or detailed structures at the interfaces so that the information can be read by the systems responsible for speech production and semantic interpretation respectively. Thus, the GoA in this case is nested as information is translated and passed along from the interior of the overall system into exterior systems interacting with the world. Following, Greco et al. (2005), we can think of this final step as the user interface after the internal computations and data type changes have taken place. Nevertheless, despite the fact that there are distinct LoAs with their own operations on their own observables at each level, they are all related to the most abstract level of syntax and the typed variables created therein.[9]

If we compare this general architecture to one of its prominent offshoots, Jackendoff (2002) PA, we get a variation on a theme. Again, we've seen some elements of the PA in Chapter 6, but here I will briefly describe the overall framework. The PA maintains the modularity and rule-bound nature of the generative architecture but rejects the central place of syntax within it (what Jackendoff calls 'syntactocentricism').

> The generative capacity of language is invested in multiple components- at the very least, autonomous generative components for phonological, syntactic, and semantic structure. Each component has its own distinctive primitives and principles of combination, and generates its own structure. (Jackendoff, 2013: 647)

Another way of putting this is that the PA maintains the autonomy of syntax without embracing its centrality. Syntactic models are then not at the most abstract level in this framework. Each linguistic system is on a similar level of abstraction. Nested GoAs tend to result in hierarchical representations of the total system. In order to escape this consequence, we might want to consider

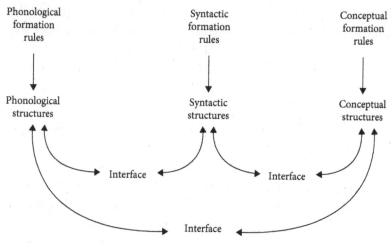

Fig. 8.5 The PA (from Jackendoff (2013, 647)).

a lattice or partially ordered set instead (see Ganascia, 2015 for more on this possibility in general).

In the PA, there are additional possibilities for interface between phonological systems and semantics which bypass syntax. This insight has been marshalled to explain languages like pidgins and creoles with limited to no syntax but potentially rich semantic and phonological structure (see Jackendoff and Wittenberg 2014). In addition, it allows for a direct correlation between prosodic contours and information structure often witnessed in topicalisation and similar phenomena. With the movement away from the centrality of syntax and with the interfaces between each independent system, the combinatorial possibilities are increased dramatically and the flexibility of the architecture is apparent.[10] The distinction between the PA and generative approaches like Minimalism can be further brought out by the difference between serial and parallel processing in computer science. The analogy is imperfect since it unclear what the processing consequences of the classical generative architecture are exactly but it helps highlight that GoAs like the PA process information in parallel with multiple processors working in tandem. This can create independent structures each with independent observables. Whereas the observables within the syntactic model of the generative picture is contained in some form at each other level which requires some sort of sequential processing (since derivations need to be created up front). In terms of the GoA, we have a situation in which "[t]he structure of a sentence is therefore an n-tuple of structures, one for each appropriate component, plus the linkages established among them by the interface components" (Jackendoff, 2013, 647). In other words, distinct LoAs with a nested GoA.

The last framework I think highlights the method of levels of abstractions well is Lexical Functional Grammar (LFG). Developed in the early 1980s by Joan Bresnan and Roland Kaplan, LFG offers perhaps the strongest case for a levels analysis of linguistic structure. Interestingly, practitioners of this account also insist, like Minimalists, that it is not a 'theory'. In this case, they prefer the term 'architecture' to 'programme':

> [T]he formal model of LFG is not a syntactic theory in the linguistic sense. Rather, it is an architecture for syntactic theory. Within this architecture, there is a wide range of possible syntactic theories and subtheories, some of which closely resemble syntactic theories within alternative architectures, and others of which differ radically from familiar approaches. (Bresnan et al., 2016, 39)

LFG also shares a certain parallelism with the PA in that it posits distinct formalisms to capture different kinds of linguistic information and then provides mappings or 'principles of correspondence' between them. The difference is that LFG splits syntax itself into distinct LoAs. The most prominently articulated levels have been f-structure (functional), c-structure (constituent and category), and a-structure (argument). But LFG also contains p-structure (phonology and prosody), i-structure (information), s-structure (syntax-semantics interface), and m-structure (morphology). We'll focus on f-structure and c-structure here for simplicity.

F-structure represents the grammatical functions such as subject and object and is supposed to be invariant between languages. The standard means of representing f-structure is as a attribute-value matrix (AVM) or an unordered set of feature-value pairs, e.g., TENSE=FUTURE, etc. Lexical items such as nouns and verbs also have a PRED feature with a unique semantic value. The f-structure shares characteristics with theta grids in Government and Binding theory and feature structures in unification-based accounts like Head-driven Phrase Structure Grammar and Sign-Based Construction Grammar. F-structure comes with a host of unique conditions such as the 'uniqueness condition' itself or the requirement that every feature has exactly one value or the 'coherence condition' which states that all argument functions must occur in the value of a local PRED feature as well as a requirement that all functions with a PRED feature also have a θ-role (thematic role).

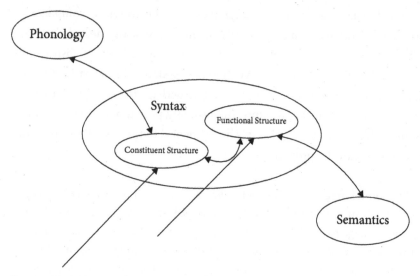

Fig. 8.6 LFG (from Debusmann, 2006).

C-structure is familiar from phrase-structure grammar. In fact, it is a modified version of X-bar theory we've seen in Chapter 7.[11] Unlike f-structure, c-structure can and does vary cross-linguistically. One of the guiding motivations for LFG was to account for nonconfigurational languages like Warlpiri in which word order is much more flexible and the phrase structure appears to be more 'flat' (cf. dependency grammar in Chapters 4 and 7). Moreover, the observables at the level of c-structure are nodes of trees not features with values as in f-structure. But syntactic representation as a whole is represented by both kinds of structure in LFG, as shown in Figure 8.6 above, with higher LoAs corresponding to each proper part of the syntactic model, i.e., c-structure to phonology and f-structure to semantics.

It's not hard to see why LFG is a clear example of the method of levels of abstraction since according to the architectural schema every sentence of natural language can be viewed from a multitude of levels of abstraction, each with their own rules, concepts, and structure.

In addition, the correspondence between c-structure and f-structure respects the nested nature of the overall GoA, since as Bresnan et al. (2016, 48) state:

Observe that the correspondence between the nodes of the c-structure and the f-structure is many-to-one. Different c-structure nodes – for example, S, VP, and V – may correspond to the same f-structure.

The mapping itself preserves information as each component of c-structure is mapped onto an f-structure which is in turn built up from the f-structures of each constituent part monotonically "which means that information is never lost or changed as part of this process; it is only added to" (Börjars, 2020, 162).

In this section, I hope to have shown the power of the method of levels of abstraction in not only characterising the relationships which hold between models of different linguistic systems but in comparing and contrasting extant architectures in which these relations prefigure. The argument and application was meant to tell us something epistemic. In other words, the account of the interfaces presented here gives us structural knowledge of the target system of natural language via the models of various theories. It might turn out that from an ontological perspective, the levels are inseparable but in terms of the joints carved for the purposes of scientific study, the method of levels of abstraction seems unavoidable to a theoretical linguist.

This picture is not unprecedented in linguistics. Ladusaw (1985) proposed a similar distinction between 'levels' and 'strata' in linguistic theory. For him, a level is one of "several systems of representation which must be based on different theoretical vocabularies," while a stratum is "one of the multiple representations of a structure when they are constructed from the same descriptive vocabulary" (Ladusaw, 1985: 15). LFG straightforwardly has two levels while generative views such as GB has only one with a number of distinct strata.

The ISR that inspired this approach is a version of ontic structural realism. I advocated for this interpretation of linguistic theory in Chapter 7 explicitly and favoured a structural approach to syntactic LRPs in Chapter 4. Furthermore, the central insight of this chapter concerned LRPs at each level of representation. What remains to be shown is that the LRPs at other levels can be viewed in an equally or relatedly structural manner. As proof of concept, the final section will attempt to show that semantics is indeed amenable to such delineation.

8.4 Semantic Metastructuralism

This section presents only a brief sketch of what a metasemantics based on ontic structural realism might look like. The idea is that what grounds semantic content or meaning is structural in nature, i.e., semantic LRPs are structural. The way I plan to argue this point is by reflecting on the nature of the most commonly used methods in natural language semantics (broadly construed), namely model theory and distributional models. Unlike the previous sections of this chapter, the point here will have more ontological import since metasemantics is ultimately the metaphysical study of semantic meaning. To set up the task, consider the three descriptive statements from recent works in the burgeoning field below:

> [P]hilosophy of language could go on to tell us *how* or *why* these symbols come to have those meanings - perhaps unearthing more basic or funda- mental facts in virtue of which such semantic states of affairs obtain. We take this second sort of inquiry to be paradigmatic, if not exhaustive, of **metasemantics**. (Burgess and Sherman 2014: 2)

> Think of metasemantics as efforts to provide metaphysical explanations of (and foundations of or groundings for) semantic facts - the fact that an expression has a certain meaning or referent. (Cappelen 2018: 57)

> The following is the organizing theoretical quest for metasemantics, before any partisan disputes set in:
> (MQ) What determines that expressions have their semantic significance? (Simchen 2018: 2)

All three statements agree that metasemantics aims to provide metaphysical accounts of how (and possibly why) linguistic symbols get their meaning or significance. In essence, we have a grounding project in which we are in pursuit of which non-semantic facts determine the semantic ones for expressions of natural language. Simchen (2018) goes on to distinguish between two kinds of answers, *productivist* and *interpretationist* respectively. The former grounds the semantic significance of expressions in terms of how they are produced or employed. Here the conditions or context of use or employment are paramount, not how they are interpreted or consumed by others. The interpretationist perspective, on the other hand, favours the consumption side in which "[t]o be endowed with significance [...] is to be interpreted or be interpretable as such" (Simchen 2018: 4).[12] Of course, this like many dichotomies is more a matter of priority (and perhaps degree) than exhaustiveness or exclusivity as both camps would presumably recognise aspects of the others' account in the total account of semantic significance. So the 'organizing theoretical quest' for metasemantics is more about what is the *primary* determinant of or ground for semantic significance.

Here, I will follow Simchen's distinction but with a methodological twist. Specifically, I will associate model-theoretic or truth-conditional semantics with interpretationist metasemantic proposals and vector-space distributional approaches with productivist views. Importantly, however, I will argue that both options put structure first ontologically and this explains why semantic LRPs are primarily modelled in terms of their structural properties and relations.

There is a natural opposition to this idea at first blush. The opposing argument would look something like this: syntax is about form or structure, and semantics is about content or 'word-world' relations. Take the concept of reference as an example of a canonical semantic relation. Individual items or SLEs receive their significance from pairing with aspects or items in the non-linguistic world. *Dog* refers to actual dogs, Barcelona refers to the Golden Retriever currently distracting me (and incidentally to a city and soccer team in Spain). What then grounds the meaning of these lexical items is something physical out there in the world (read: mind-independent). Larger expressions get their significance via disquotational schemata with similar compositional world-boundedness.

Importantly, the opposition is not tracking semantic externalism versus internalism. Although, historically, internalists like Chomsky have also been neo-structuralists about syntax, internalism could also favour a view of semantics as content-driven, only the content is conceptual and internal on their views.[13] To adapt a famous example from Chomsky, to what does 'Barcelona' refer? The geographical area in northeastern Spain, the people inhabiting the region, the buildings and architecture or my ever more successfully distracting dog?[14] Chomsky's own example is below.

> Referring to London, we can be talking about a location or area, people who sometimes live there, the air above it (but not too high), buildings, institutions, etc., in various combinations (as in London is so unhappy, ugly, and polluted that it should be destroyed and rebuilt 100 miles away, still being the same city). Such terms as London are used to talk about the actual world, but there neither are nor are believed to be things-in-the-world with the properties of the intricate modes of reference that a city name encapsulates. (Chomsky 2000a: 37)

Internalists, following the logic above, reject the standard definition of reference as involving *externalia* or non-mental states. However, they do not endorse a conceptual view of content in terms of ideas or propositions in the mind either. Rather semantics is in the province of syntax in that the proper task of semantics is to provide information to the SM and CM interface systems which can enable the individual language user to employ the term 'London' for communicating about various aspects of the language-external world (Chomsky 2003: 294–295).

The view I am going to suggest here is not incompatible with internalism or externalism but quite orthogonal to that debate. The structures that ground our meaningful use of words and expressions might indeed be mental instructions to the interfaces or conventions grounded in linguistic practice. It is also not unrelated to the ontological account of words I presented in Chapter 6. There, I claimed that the ontology of words is determined by their places in linguistic structures (in turn, found in biological systems). Similarly, the claim here is that *the meaning of a term is determined by its place in a larger linguistic system.* In other words, we can think of semantic metastructuralism as a form of holism.

Think for a moment of an eco-system. One could describe an organism, such as a particular succulent or small animal, individually. But there is a reason that nature documentaries often choose to represent the entire region with its various interconnections and co-dependencies. Understanding the behaviour of a Finch in the Galapagos Islands is impoverished at best without an understanding of the dynamics of the surrounding environment, predation and competition. This is not just an epistemic point, the Finch's existence (its height, weight, colour and behaviour) is determined these factors (as Darwin discovered). In other words, the meaning of Finch (in that context at least) depends on the meanings of other creatures and features of the environment.

In this vein, there are two famous principles in semantics attributed to Frege. The first is 'the context principle' which states that we should never "ask for the meaning of a word in isolation, but only in the context of a sentence" (Frege 1953: x). This principle dovetails well with the claim I make above. In a sense, my claim about meaning can be considered a kind of *global* context principle since it extends the idea that meanings are determined by larger contexts, in this case the natural language defined as a complex system from Chapter 5. On the other hand, the principle of compositionality which is the bedrock of formal semantics can be seen as making an opposing kind of claim. First assign meanings to individual parts of an expression, then build the complex meaning up from those (and their syntactic combination). There's a huge literature on defending each principle and whether or not they are compatible that will take us too far afield here.[15] Instead, I think we should follow Linnebo (2008b) in a particularly relevant means of reconciliation between the two principles.

He claims that in order to appreciate the compatibility of the context principle and compositionality, we need to invoke the semantics-metasemantics distinction. Specifically, he argues that the principle of compositionality belongs at the level of semantics which is "concerned with how the semantic value of a complex expression depends upon the semantic values of its various simple constituents" (Linnebo, 2008b: 24). Metasemantics, as we have described it, is the investigation of "an expression's having the semantic properties that it happens to have, such as its semantic structure and its semantic value" (Linnebo, 2008b: 24). Thus, the context principle, and *a fortiori* my global version, is a metasemantic hypothesis.

So if we are setting ourselves the task of describing the semantic value of a particular sentence or expression, compositionality will guide us as to how to combine the meanings of each part (usually by means of type-theoretic combinations and functional application) in order to get to the value of the whole. This is largely methodological. But when we are asking deeper questions about the nature of meaning itself and why something like compositionality works, then we have to appreciate the whole structure first and then understand individual meanings in terms of it. Structures are thus prior metasemantically to individual units of meaning.

This is all to spell out the claim for which I am pushing and the level at which it is pitched. I still haven't provided an argument for it. The remaining part of this section is a sketch of the argument for *semantic metastructuralism*.

Firstly, why 'metastructural' and not 'metasemantic'? The answer is not very meta. Basically, I don't think we need anything as strong as the entire language to be able to ground meanings since it is unclear what it even means to know the entire language. Do English speakers know English in some total system manner? This seems unlikely, we probably only have acquaintance with local instantiations of the total system. Not only this but although the English system of South Africa is related to the North American and British systems, it is identical to neither. For example, South African English is invariably affected by massive language contact from the earlier days of colonial settlement and the Khoisan and the Nguni languages indigenous to the region as well the variants of foreign languages like the Dutch-derived Afrikaans. So we only need subsystems of the total system, or *metastructures*, to ground our meanings and these will be determined by the (diachronic) linguistic environment in which they occur.[16]

More importantly, I think metastructures are the ontological building blocks of meaning due to the kinds of tools we predominantly use in its investigation. On the interpretationist side, model theory has been widely adopted since Montague (1970) as a mathematically precise way of determining semantic structure. Models are basically interpreting structures. For instance, very informally, we can define structures in terms of *signatures* in model theory such that a signature is just a tuple consisting a sets of constant symbols, function symbols, and relations (the latter two sets usually come with arity functions). The structure S defined on this signature set-theoretically consists of some domain, M, that assigns objects to the constants, functions and relations of the signature. If we want to define different structures, we either change the domain or the assignments (or both). A *model* is just a structure that makes certain sentences true or false. So model-theoretic truth amounts to truth under a certain interpretation.[17]

In model-theoretic semantics for natural language, each expression is associated with a particular set-theoretic object, e.g names and ordinary objects are $< e >$ type, intransitive verbs are functions from these entities to truth values $< e, t >$ and sentences are $< t >$ and so on.[18] In order to limit confusion and name these various set-theoretic objects, λ-notation (with variables) from formal logic was introduced. Thus, a simple sentence like Jeff works becomes $\lambda x(x \text{ works})(\textit{Jeff})$ where we replace the name *Jeff* with a variable x, add the name as an argument, and attach the λ-operator to bind the variable with the result that the λ-formula now represents a truth-conditional function. The puzzle is then how to define individual items and expressions in such a way as they combine to form subexpressions and finally whole sentences. The principle of compositionality guides this process in semantic analysis.

It is imperative to note that the particular meanings *qua* set-theoretic objects are not individually ontologically committing. It's the combinatorics that really matters. Jacobson spells this point out nicely with reference to the use of lambda notation:

> [T]he language used to name model-theoretic objects is of no consequence for the theory and is only a tool for the linguist (the grammar itself simply pairs expressions with model-theoretic objects). But once the objects get sufficiently complex (e.g., functions with functions as their domains as well as co-domains) it is useful to have a simple way to write out these meanings without excruciating and/or ambiguous prose, and the lambda calculus allows us to do just this [. . .] a reminder that these name actual model-theoretic objects and that the formulas have no theoretical significance. (2014: 134)

Basically, model-theory just models meaning in terms of interpreting structures. Therefore, it is a structural enterprise much like I have described syntax in the book. Delving into the lexicon or lexical semantics might involve more than just structure but standard model-theoretic semantics seems largely about the kinds of mathematical structures we associate with aspects of the grammar.

Each syntactic structure is associated with a semantic counterpart which contributes to the complex meaning of whole expressions. But as *per* the global context principle, from a metasemantic perspective, the components of these formal objects must map onto some structure in the external world (or internal state of the language user). Here the explanation goes in the opposite direction, top-down or holistically if you will.[19] These are the metastructures, which are proper parts of larger linguistic systems. The interpretationist metasemantic perspective then provides evidence for semantic metastructuralism.[20]

What about the productivist side of metasemantics? Well, the best example of this approach to semantic methodology is in vector-space distributional models. Instead of a meaning or semantic value being defined by the association of a word or expression with a discrete set-theoretic object, semantic values are represented as vectors or more abstractly meanings of SLEs are modelled "as points or regions in some "meaning space"" (Erk 2020: 71). Here, statistical collocational data provides the background for representing semantic structure. A simple case is 'count-based spaces' in which a vector is a sequence of numerical values used to measure the meaning of a word in terms of the words which co-occur with it across multiple contexts. The underlying assumption is that words with the same or similar meanings occur within similar contexts. By representing a word as a matrix of similarity scores within these contexts, we can thus we get to its approximate meaning. In essence, we are looking for semantic patterns in the data and identifying meanings with places or 'spaces' in these patterns. So far so structural.

Nevertheless, the explanation we are in search of in this section is metasemantic in nature and for the longest time, distributional approaches were relegated to practical computational purposes not to be considered seriously in theoretical contexts. Erk thinks the tide is turning.[21] She distinguishes between semantic spaces 'in practice' and 'abstract semantic spaces'. I think the latter are at the level of metasemantics for the following reason.

> By abstract semantic spaces, I mean approaches that build on the idea of a space – meaning space, or conceptual space – as a way of representing meaning. There is no uniform definition of an abstract semantic space, rather a collection of approaches that differ in which properties of practical semantic spaces they adopt, and that differ in the phenomena they most care about. (Erk 2020: 72)

If there are different practical approaches to demarcating the abstract notion of a semantic space, then those approaches seem to be on the semantic level while the abstract space is at the metasemantic level. So for instance, the count-based approach I described briefly above is a semantic level procedure for mining data for approximations of meaning based on similarity of context. But the corresponding metasemantic level admits various grounding possibilities including Gärdenfors' (2000) conceptual spaces or even a semantic space represented as a model structure as in the model-theoretic approach (Herbelot and Copestake 2013) where "the 'vector' associated with a predicate symbol would consist of all the individuals (or tuples of individuals) of which the predicate symbol is true – its extension" (Erk 2020: 79).[22] Either way, viewing meanings as places in abstract semantic spaces incorporates tools in search of structural patterns. What grounds these meanings is the structures or semantic webs so identified. Again, internalist (e.g., Gardenfors' conceptual spaces) or externalist options in terms of corpora of actual linguistic communication or public language abound.[23] Abstract semantic spaces are semantic metastructures in the sense I have been describing because they ground meanings as structural positions in semantic patterns of linguistic systems.

In this section, I've only provided a sketch of one possible extension of structural realist ideas to semantics. I've done so by first aligning semantic theoretical approaches with Simchen's metasemantic positions of interpretationism and productivism, then I argued that both options can be exemplified by current trends in semantics, model-theoretic, and distributional, respectively. Finally I suggested that in either case, we are identifying meanings at the metasemantic level with certain kinds of structures. There is certainly more to be said here but this sketch will have to suffice for present purposes.

The chapter as a whole had the express aim of providing a framework for discussing the interfaces between linguistic systems and understanding and comparing how different extant architectures have attempted to capture this interplay. The next chapter broadens our perspective once more, this time with the aim of understanding the place of linguistics within the larger cognitive scientific project.

9

Language and Cognitive Science

An Arranged Marriage

9.1 The Dilemma

Language has had a central place in the emerging field of cognitive science since the latter's inception in the late 1950s. The centrality of language has owed its position largely to the influence and success of generative linguistics during that period. However, in part due to challenges to the generative paradigm from within linguistics, language is becoming more peripheral in the so-called Second Generation Cognitive Science (Sinha 2010) of the early 2000s with cognitive psychology taking centre stage. Not only this, but the field of cognitive science is becoming more disunified. I think there is a connection between these two occurrences. Therefore, in this final chapter, I have a relatively modest argument towards the following central insight.

CENTRAL INSIGHT VIII: LINGUISTIC THEORY, VIEWED STRUCTURALLY, DOVETAILS WITH SOME OF THE MODELLING PRACTICES OF THE COGNITIVE SCIENCES CONSTRUED IN TERMS OF STRUCTURAL REALISM.

The basic argument is as follows: after the rise of connectionism and 4E approaches to cognition, linguistics became the 'odd discipline out' in the cognitive sciences, in part because it retained a style of theorising characteristic of older computationalist theories of mind. However, viewed from the purview of the present structural realist framework convergence of modelling strategies and ontological commitment can be aligned. This latter realisation will facilitate easier integration of results from linguistics with results from other cognitive scientific disciplines.

Thus, in Section 9.2.1, I briefly discuss the history of mentalistic linguistics and in Section 9.2.2 do the same for cognitive science. In Section 9.3, I analyse different cognitive architectures for the field and suggest an intersectional model based in part on biological systems theory. Lastly in Sections 9.3.2 and 9.4, I make a case for why language should not be so easily discarded

Language, Science, and Structure. Ryan M. Nefdt, Oxford University Press.
© Oxford University Press 2023.
DOI: 10.1093/oso/9780197653098.003.0009

as central to the cognitive sciences by rejecting the claim that linguistics is isolationist in methodology before discussing how a notion of cognitive structure in terms of structural realism might support the architectural picture under discussion.

9.2 The Study of Mind in Language

9.2.1 A Short History of Mentalism in Linguistics

The standard story concerning the relatively recent history of generative or formal linguistics starts with the idea that it emerged out of the limitations of its predecessors, the motley assortment of views characterised under the umbrella term 'structural linguistics'. Structuralism, on this view, aimed to produce taxomies of individual languages and their linguistic structures. Certain advocates, such as Bloomfield, were sceptical of talk of the mental within linguistics.

> Non-linguists (unless they happen to be physicalists) constantly forget that a speaker is making noise, and credit him, instead, with the possession of impalpable 'ideas'. It remains for linguists to show, in detail, that the speaker has no 'ideas', and that the noise is sufficient. (Bloomfield 1936: 23)

I am not planning to disrupt that narrative here (for that see Joseph 1999, Matthews 2001). Rather I plan to trace a different aspect of the emergence of the study of language under the generative paradigm, namely the understanding of linguistic structure as a reflection of cognitive structure. Even if there was some formal continuity between linguistic structuralism and generativism (Chapter 7), there does seem to have been a marked shift with the advent of the mentalist ontology.

The formal mathematical tradition in linguistics can be argued to have been established by Chomsky (1956) and (1957).[1] Mentalism in linguistics has a slightly different trajectory which can be traced back to Chomsky's (1959) review of B.F. Skinner's *Verbal Behavior* and picked up again in Chomsky (1965).

We have seen mentalism in Chapter 2 from an ontological perspective. Roughly it is the view that language can be defined as a mental state or a mental abstraction from a brain-state. In other words, language is a cognitive system.

Specifically, generative linguists claim that the object of study is a particular internalised and idealised state of a language user's mind.

> Generative grammar adopted a very different standpoint. From this point of view, expressions and their properties are just data, standing alongside of other data, to be interpreted as evidence for determining the I-language, the real object of inquiry. (Chomsky 1991: 12)

As we have seen before, an *I-language* is a rule-bound internalised conception of the structural descriptions associated with grammatical expressions of a language. It is contrasted with a E-language or external language often characterised as a set of behavioural dispositions of individual linguistic communities and often marked in terms of sociolinguistic boundaries. Chomskians have traditionally remonstrated against a possible scientific understanding of language emerging from this latter perspective. In Chapter 5, I provided a novel framework for uniting these two components of natural language within a naturalistic paradigm.

At the time of the Classical Cognitive Scientific revolution of the 1950s,[2] language proved an essential platform for the study of the mind. The experimental clarity of Behaviourism made it highly influential in American psychology and the strictures of logical positivism reigned in any ontological expansions beyond a parochial conception of parsimony. Psychology was firmly under the sway of the former and philosophy still within the grips of the latter.[3] Thus, neither were in a position to lead a counter-revolution and liberate the concept of mind from the intellectual climate of the time.

Chomsky's (1959) critique of the radical behaviourist proposal of Skinner (1957) on language challenged the stronghold that such accounts had on the study of language. Unlike animal calls, which arguably could be analysed in terms of stimulus and response, human communication is stimulus independent. Following an insight from Wilhelm von Humboldt on the infinite capacity for linguistic expression based on finite resources, Chomsky extended the mathematics of the time to capture linguistic unboundedness in terms of generative grammars.[4] Again, concepts beyond the reach of radical behaviourism.

If language is such an essential component of what makes us human and its nature can be identified with a kind of mental competence, then linguistics could be the field to banish resistance to the explicit study of the mind. This precise claim was made in Chomsky (1965), which reset the linguistic agenda firmly in favour of mentalism. The now infamous passage from *Aspects of the Theory of Syntax* reads:

> Linguistic theory is concerned primarily with an ideal speaker-listener, in a completely homogeneous speech-community, who knows its (the speech community's) language perfectly and is unaffected by such grammatically irrelevant conditions as memory limitations, distractions, shifts of attention and interest, and errors (random or characteristic) in applying his knowledge of this language in actual performance. (Chomsky 1965: 4)

Not only this, but a language on this view is to be identified with mental competence in that language (later *I-language*). As we've seen, linguistics redefined itself as a subset of psychology and in so doing released a respectable mentalism back into the latter. "If scientific psychology were to succeed, mentalistic concepts would have to integrate and explain the behavioral data" (Miller 2003: 142). The story goes that talk of 'the mind' was moving past the occult into mainstream science. Linguists started to investigate concepts of modularity (under the 'Government and Binding' banner), developmental psychology (within language acquisition studies) and the cognitive substrate underlying all human languages (via the Universal Grammar postulate).[5] Pylyshyn described linguistics at the time in the following way:

> [D]espite the uncertainties, none of us doubted that what was at stake in all such claims was nothing less than an empirical hypothesis about *how things really were inside the head of a human cognizer.* (1991: 232)

The revolution in linguistics therefore played an important role in the establishment of the cognitive sciences. However, it was not the only factor that led to the field emerging as a multidisciplinary enterprise. It is on to this complicated history that we now move.

9.2.2 The Inception of Cognitive Science

Bolstered by the academic successes of formal linguistics across North America (and the funding initiatives of the Sloan Foundation), the rigorous study of cognition could finally emerge as a distinct and interdisciplinary field of inquiry.

The Cognitive Science Society was officially founded in 1978 but George Miller, one of the original pioneers, traces the conception of the field to a particular day in 1956.

> I date the moment of conception of cognitive science as 11 September, 1956, the second day of a symposium organized by the 'Special Interest Group in Information Theory' at the Massachusetts Institute of Technology. At the time of course, no one realized that something special had happened so no one thought that it needed a name; that came much later. (Miller 2003: 142)[6]

This symposium brought together experts in the diverse fields of Artificial Intelligence (AI), psychology, early cognitive neuroscience, information theory and of course linguistics. Since the inaugural conference, certain disciplines have moved out of prominence while others such as philosophy have entered into the field.

> Computer science and psychology have played a strong role throughout. Neuroscience initially was strong, but in the years immediately after the 1956 conference its role declined as that of linguistics dramatically increased. (Bechtel et al. 2001: 2154)

There were a number of early disciplinary influences which ultimately culminated in the emergence of cognitive science in the mid-twentieth century. Some of these influences came from outside of the North American academic milieu while others came from the perceived limitations of approaches available at the time such as informational theory and behaviourism. In the following subsections, I will detail these various developments.

9.2.2.1 Influences from Without

Although 'cognition' and 'the mental' might have been unpopular terms within the scientific lexicon in North America in the 1950s, there were theorists in various fields outside of the United States who embraced mentalistic terminology. I will briefly survey some of the salient figures which prefigured the development of cognitive science.

In Germany, the work on the psychology of language (or *Sprachpsychologie*) of Wundt and the Gestalt psychology of Koffka and others did not shy away from mental categories or 'Gestalt' qualities of perception. Importantly, the idea of *Gestalten* was developed in direct opposition to basic behavioural concepts. "Koffka took his argument a stage further in 1915, arguing for a revision of the concept of "stimulus", which should no longer be seen as a pattern of excitation, but as referring to whole, real objects, in relation to an actively behaving organism" (Sinha 2010: 1278). The Gestalt principles identified a distinctly cognitive dimension of perception not directly reducible to behavioural dispositions. Their position was an interesting mixture of a commitment to physicalism and the autonomy of the mind and its study. Epstein and Hatfield (1994) describe Gestalt psychology as at once committed to 'phenomenal realism' taking mental experience and internal states as real and objective and 'programmatic reductionism' which aims all eventual complete explanation at the physiological level.

Another major influence from continental Europe was the genetic epistemo-logical approach of Piaget. The latter was, much like behaviourism, explicitly experimental but unlike behaviourism did not obviate talk of the mental or consciousness. In fact, the so-called 'epigenetic' account of development or growth of an organism advocated development as conditioned both internally and externally. For practitioners of this view, understanding knowledge acqui-sition involved essentially understanding the nature of subject and object and the development of intelligence from infancy. The view set itself apart from psychology at the time by rejecting the claim that knowledge involved some passive reception of external factors and by positing a distinctive aspect of 'construction'. The model of stimulus and response was thereby modified in terms of the biological analogy of *assimilation* and *association*.

> Indeed, no behavior, even if it is new to the individual, constitutes an absolute beginning. It is always grafted onto previous schemes and therefore amounts to assimilating new elements to already constructed structures (innate, as reflexes are, or previously acquired) [...] Similarly, in the field of behavior we shall call accommodation any modification of an assimilatory scheme or structure by the elements it assimilates. (Piaget 1976: 18)

Of course, related to this, in the United Kingdom around a similar time, the concept of cognitive schemas was being investigated by Sir Frederic Bartlett. Bartlett criticised the allegedly vacuous accounts of schemas according to which they were merely "storehouses of sensory impressions" and advocated rather that "[s]chemas, are, we are told, living, constantly developing, affected by every bit of incoming sensational experience of a given kind" (1932: 201). Schemas thus involved complex cognitive operations not unlike the accommodation notion of Piaget above. His reconstructive account of memory emphasised the active element in the production of reflective narratives of past events not by merely recording actual features of those events but by recreating and reconstructing them in terms of their meaning and inferences therein involved ('the gist' of the story). The mind, then, actively combines records of actual details with the knowledge contained in pre-existing schemas. This is ostensibly far from a 'black-box' notion of the mind.

The above, of course, is only a snapshot of some of the influential research which was being conducted outside of the cognitive programme being developed in North America in the 1950s.[7] Sinha (2010) claims many of the above influences skipped a generation and rather informed contemporary cognitive linguistics and second generative cognitive science but it was clear that early cognitive scientists were well aware of these developments and in part inspired by them. As Miller notes, reflecting on the time of the Classical Cognitive Revolution:

> In Cambridge, UK, Sir Frederic Bartlett's work on memory and thinking had remained unaffected by behaviorism. In Geneva, Jean Piaget's insights insights into the minds of children inspired a small army of followers. And in Moscow, A.R. Luria was one of the first to see the brain and mind as a whole...Whenever we doubted ourselves we thought of such people and took courage from their accomplishments. (2003: 142)

9.2.2.2 Influences from Within

There were a number of direct causes which led to the formation of a distinct field designed to explore cognition. Two main types of causes stand out among the others, namely what I call *development from limitation* and *development from connection*. The first type is easily seen in the case of behaviourism. Its methodology was restrictive and thereby its results were limited as shown by Chomsky (1959) for the case of language.

Another source of frustration was related to the nature of information theory at the time. The place of Claude Shannon's work on making the notion of *information* mathematically precise and ushering in the digital era is unquestionable. Shannon's implementation of Boolean logic by means of relays and switches of electrical circuits laid the foundation for modern computers. As a theory of psychology, however, it had its drawbacks. For instance, the finite Markov processes Shannon used to analyse natural language opened the door for Chomsky's transformational grammar approach which could handle syntactic structures the former could not (shown in Chomsky 1956). Markov processes of the time mapped well onto behavioural output but for a theory of mental manipulation of grammatical rules it was lacking.[8]

In terms of development by connection, there were a number of emerging fields besides linguistics that had direct application to human cognition. In the early days, the artificial intelligence of Minksy and McCarthy played a central role in cognitive science. Similarly, Newell and Simon's application of computer science to cognitive processes and problem solving set the stage for cognitive psychology to merge with information processing and computer science. More and more advanced techniques were being developed to study the mental realm and the field was primed for the advent of a new interdisciplinary programme in the study of mind.

AI and cybernetics certainly provided the conceptual tools and formal rigour required for modelling mental processes. However, although AI was initially a major component of cognitive science, it waned in subsequent years. Part of the reason for this was that AI moved from a model of human imitation inspired by Turing to one based on amplifying human abilities.[9] "In AI, it is perfectly legitimate to use methods that have no relation to human thinking in order to attain functional goals" (Chipman 2017: 7). Linguistics provided a further explanatory dimension in its grammatical analysis of complex syntactic patterns. But it was the neuroscience of the time that connected the mind to the brain in interesting ways. The early interdisciplinary work of McCulloch and Pitts on neural networks inspired the later connectionist models of the mind (in the Second Generation Cognitive Science) but the identification of specific areas of the brain and their functions, such as Broca's area associated with language, become of utmost importance. Studies on aphasia found their way into linguistics textbooks and defects in mental competence in language were associated with neurophysiological causes cementing the psychological inter-pretation of language studies. Even the competence-performance distinction was in part motivated by neurophysical evidence. As Stabler notes in defence of the distinction,

> The linguistic idealization is apparently grounded in the empirical assump-tion that the mechanisms responsible for determining how phrases are formed in human languages are relatively independent of those involved in determining memory limitations, mistakes, attention shifts, and so on. (2011: 70)

According to Miller, psychology, computer science and linguistics were to be the core fields within cognitive science. In addition, "[o]ne of the central inspirations for cognitive science was the development of computa-tional models of cognitive performance" (Bechtel et al. 2001: 2155). These two aspects culminated into one of the most influential ideas of the entire movement, later embraced both by linguistics and philosophy, namely the computational theory of mind. However, the emphasis on performance has also led to a challenge to the competence model of generative grammar mentioned in Section 7.2 and the rise of psycholinguistics within cognitive science (more on this in Section 7.5). In the following section we look at the computational theory of mind and its ramifications for the Classical Cognitive Revolution.

9.2.3 Language, Representationalism, Computationalism

What linguistics showed possible and what AI and computer science successfully modelled was the idea that cognition could be viewed in terms of the manipulation of mental representations *qua* computational states. This is the computational theory of mind. The notion that the mind can be understood as a symbol manipulator much like a modern day computer, where the mind is at the software level and the hardware is the brain. Two aspects of this view are important for present purposes, computationalism and representationalism. We'll deal with each in turn.

The classical version of the thesis proposed that the mind can be understood as a Turing machine of some sort. There are a number of versions of this claim, computational neuroscience and even connectionism (which diverges considerably from the Turing model) can be viewed as such. The basic idea which underlies such views is that the mind is a computational system. Chalmers (2011: 232) identifies two aspects of this claim.

> First, underlying the belief in the possibility of artificial intelligence there is a thesis of *computational sufficiency*, stating that the right kind of computational structure suffices for the possession of a mind, and for the possession of a wide variety of mental properties. Second, facilitating the progress of cognitive science more generally there is a thesis of *computational explanation*, stating that computation provides a general framework for the explanation of cognitive processes and of behavior.

The latter is essential for understanding computationalism and cognitive science in general. The idea of generative grammar assumes this position in defining grammars which operate on syntactic categories in generating all the well-formed expressions of the language. It is a computational system, a symbol manipulation device which operates on mental items. Linguists posited that this computational device was isolated from the rest of the cognitive system. The autonomy of syntax was one explanatory means of isolating the computational aspect of mental language competence. The idea, initially proposed by Chomsky (1957), is that a grammar (or generative grammar) of a language L constitutes a scientific theory of an internalised rule system for generating all and only the grammatical sentences of L. A speaker 'knows' or 'cognises' this system of rules and produces judgements or linguistic intuitions based on this knowledge of the grammar.

But generative linguistics embraces a particular brand of computationalism, namely the representational theory of mind (RTM). RTM is a version of CTM which takes symbolic manipulation to the level of mental representations of a specific sort. One radical expression of this view is captured by the work of Jerry Fodor (1975, 1981, 1987) and the *Language of Thought Hypothesis* (LOTH) which held that there is a mental language or *Mentalese* upon which the computational or Turing system operates. This language is compositional, productive and systematic according its advocates. Chomskians do not generally adhere to the strong claim of LOTH due to its semantic nature. They do endorse representationalism of a slightly different order though.

The rules of the language, syntactic rules, are internally represented by speakers of the language. In characteristic fashion, Chomsky states, "there can be little doubt that knowing a language involves internal representation of a generative procedure" (1991: 9).[10] Thus, language and generative linguistics provided a viable case for computationalism and representationalism in the rest of the cognitive sciences. All of these properties, however, were challenged in the Second Generative of Cognitive Science. It will nevertheless be my claim that language remains to be a central feature of the field despite the potential paradigm shift.

9.3 Intersection, Integration, and Architecture

The philosophy of cognitive science identifies a distinctively methodological worry within the field related to its own architecture. "Cognitive science is, of course, a multidisciplinary field of research, but there remains enormous disagreement regarding the relevance and precise role of these various disciplines" (Samuels et al. 2012: 10). For example, so far, I have been using the terms 'cognitive science' and 'cognitive sciences' interchangeably. But there is a potentially important distinction to be had here. The distinction essentially tracks the difference between cognitive science as a discipline versus the cognitive sciences as a pursuit. It's a matter of union versus intersection.

> Thus, one possible definition of *cognitive science* would include all work in these contributing fields, and everyone working in these fields might be deemed *cognitive scientists*. That is the *union* definition of cognitive science. An alternative is the *intersection* definition of cognitive science. To qualify as cognitive science under this definition, the work must draw on two or more of the contributing fields. (Chipman 2017: 1)

The union approach is multifarious and can include even more fields than has been mentioned here such as economics and sociology. However, it seems like a bad place to start in defining a field since all of the various parts can have little to nothing to do with one another, hence the more suited term 'cognitive sciences'. In addition, they would be permitted to pursue independent goals. Under this interpretation, AI's movement beyond modelling human thinking would not remove it necessarily from cognitive science, a counter-intuitive situation. This architecture allows for connection but does not require it. Furthermore, a union does not produce an ordered set and the core fields would have no claim to special status. The general position is represented in the Figure 9.1.

On the intersectional approach, the characterisation of cognitive science aims at finding the core approach(es) within various disciplines aimed at cognition. This strategy seems to be more in keeping with the original conception of cognitive science. The alternative architecture is one of homing

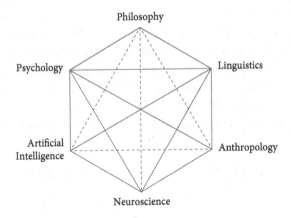

Fig. 9.1 Gardner (1985), *The Mind's New Science.*

in on the mental and dividing core disciplines from peripheral ones more starkly. The debate then becomes about which disciplines are more beneficial to have at the core than others. In many ways, the *Second Generation Cognitive Science* makes a strong case for situating psychology as the central field and relegating linguistics to the periphery. This move makes some sense as psychology is precisely the study of the mental processes and it is by historical accident (mainly its association with behaviourism) that resulted in it not claiming that space in the Classical Revolution. However, psychology does not intersect with all of the relevant fields within the cognitive sciences and advocates might thus favour a union approach for its centrality.

One could of course object that the choice of architecture or core versus periphery makes little difference to the study of mind. I would caution against such a position. One of the main reasons for the development of cognitive science in the first place was that the study of the mind was too complex and broad for a single methodology to capture. An interdisciplinary approach was needed. However, if one understands such an approach as a union of different fields, then shared insights will be hard to come by as theorists continue to operate largely in silos. The resulting field would be interdisciplinary only in name. In order for related goals to be pursued surely a unified approach is necessary. To borrow a metaphor, theorists need to speak each others' languages or at least dialects of the same language. As Sinha (2010) admits, "[t]here can, however, be little doubt that contemporary cognitive science is much less consensual in its fundamental assumptions than was the case a quarter of a century ago" (1266). But some consensus is necessary for the progress of a paradigm, at least for 'normal science' (in the Kuhnian sense) to continue. And any discipline or concept which could unify a subject would be advantageous for precisely this reason. The alternative picture suggested by these considerations is outlined in the Figure 9.2.

In what follows, I hope to cast some doubt on the claim that psychology should replace language as a core of cognitive science by arguing that language and linguistics better serves the intersectional approach to cognitive science. In Section 9.4, I will go further to suggest that the intersectional approach on the metatheoretical level dovetails with a structural realist understanding of cognitive science based on neuro- and computational cognitive science.

9.3.1 Lessons from Systems Biology, Again

So far the motivation for an intersectional approach to cognitive science has been quite high level. In this subsection I will discuss a case study from

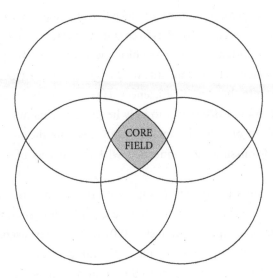

Fig. 9.2 Intersectional Approach.

molecular systems biology as an example of a fruitful intersectional or rather integrational endeavour. Here I will assume that the mechanism behind the intersectional approach is that of *integration* where the latter is understood as the combination of cross-disciplinary methods, tools, and models aimed towards a particular or generalised target. In recent times, the life sciences have been confronted with a deluge of molecular data fuelled in part by high-throughput data collection projects in genome sequencing and systems biology. In fact, "[s]ystems biology, the successor science to standard [data-driven] genomics, is usually described as the field that has arisen out a perceived need for more integration, particularly of data and methods but also disciplines" (O'Malley and Soyer 2012: 59). The issue is that massive data collection leads to the increased appeal of bottom-up theorising but there are genuine discoveries to be had by means of hypothesis testing and mathematical modelling. One methodological approach does not seem viable for the unification of the field nor do isolated methodologies always lead to novel discoveries. O'Malley and Soyer (2012) pick out three cases of the successful integration in biology, vesicle transport in cell biology, noise biology, and evolutionary systems biology. I want to focus on the last of these as I think it provides the most relevant case for the cognitive sciences.

We have already encountered systems biology in Chapter 5. So I won't redescribe the field here. Basically, systems biology takes individual organisms to be the less natural abstractions from the true individuals of the biological sciences, namely microbial communities in their natural environments. *Prima facie,* the move to the 4E approaches to cognitive science resembles the move to systems from individual organisms in biology. Most of the 4E approaches take environmental factors to be constitutive of cognition and advocate integrating social sciences into the cognitive sciences. However, as previously mentioned the general 4E framework seems to only be united by its mutual rejection of computationalism or cognitivism. I'll briefly describe the main components below before dismissing the connection with the success of systems biology here.

The first E refers to *embodied* cognition. This is the view that cognitive processes are *dependent* in part on 'extracranial' factors such as the kind of body the cogniser occupies. Newen et al. (2018) make a few useful distinctions. Firstly, 'extracranial' can mean outside of the mind so as to include the body or it could extend beyond the body to the environment. Secondly, *dependence* can be understood either *causally* or *constitutively.* If the latter, the view is much stronger. Indeed, some form of embodiment is assumed across the 4E catalogue but each view comes with its own sets of commitments and aims. Embodiment not only constrains an organism but limits their phenomenological horizons, i.e., literally shapes what the world looks like for them. *Embedded* cognition expands on this view by incorporating the natural and social environment in the understanding of human cognitive processes. Thus concepts are *situated* in that they cannot be separated artificially from the environment in which they are found, i.e., prior knowledge figures inseparably. And concepts are *embedded* when their situatedness is internalised. For example, if a cognitive process like counting is developed within, say, an agricultural environment in order to save time or effort on farming tasks. This process can be internalised for use outside of the specific environment in which it developed. This E is similarly to some of the core assumptions of systems biology. The *extended* claim allows for the inclusion or extension of cognitive processes into external objects or artificial devices used to perform cognitive tasks. Standard examples are eye glasses for the visually impaired or more starkly a blind person's stick used to 'extend' the perceptual domain in the absence of sight (Martin 1993, De Vignemont 2018).

In their foundational work, Clark and Chalmers (1998) aim to use extended cognition to reflect on not only the idea of mind or cognition but also concepts such as *BELIEF.* They propose what they call 'active externalism' to establish the connection between an organism's environment and its cognitive function.

[T]he human organism is linked with an external entity in a two-way interaction, creating a *coupled system* that can be seen as a cognitive system in its own right. All the components in the system play an active causal role, and they jointly govern behaviour in the same sort of way that cognition usually does. If we remove the external component the system's behavioural competence will drop, just as it would if we removed part of its brain. Our thesis is that this sort of coupled process counts equally well as a cognitive process, whether or not it is wholly in the head (Clark and Chalmers 1998: 9).

Thus, when we use calculators to extend our arithmetic abilities, we are joined with the external object in a sort of cognitive union. Clark and Chalmers offer many examples of this phenomenon.

Lastly, *enactivism* takes yet another angle in requiring both embodiment and embeddedness by adding goal-direction into the mix. This view adds agency to the overarching framework in that human agents co-create their world and that cognition is partly based on a cogniser's actions. Valera et al. (1991: 9) describe the approach as one that "emphasise(s) the growing conviction that cognition is not the representation of a given world by a pregiven mind but is rather the enactment of a world and a mind". This 'enactment' is based in turn on the kinds of actions the subject performs in the world. I can't go into the many vicissitudes of this approach or the overarching 4E take on cognition here. But there are two relevant disanalogies in terms of integration. The first has already been mentioned. Despite catchy terms like '4E' there is little unity across this framework. For instance, it is not clear if enactivism requires embodiment. The second issue is that the overarching paradigm still focuses on the individual as the nexus of cognition unlike the systems approach or the structural realism for linguistic theory I have been arguing for. In fact, in the linguistic case, there would be more focus on individual idiolects with their *sui generis* embodiment and situatedness in particular communities. This is problematic for any unified theory of cognition. Integration is therefore an unlikely tool for such research. This is not to say that the 4E approaches aren't relevant to cognitive science or should not be incorporated but rather that they should be one or many of the multiple models used in the service of the unifying task.

Returning to the discussion of integration in systems biology. The paradigm shift has been largely synchronic. "[S]ystems biology tries to understand life as it is now [...] [it] does not yet aim at explaining the evolution of biological systems" (Boogerd et al. 2007: 325). When an integration of various mathematically and experimental methods such as signalling and genetic circuitry were applied to the functional synchronic field, evolutionary systems biology emerged as a new diachronic study of genes and organisms. By integrating various fields and methods, a new line of inquiry was formed, one which aims at answering complex questions concerning the origins of complex organisms in the life sciences.

> The very act of integrating evolutionary knowledge and practice within the realm of molecular systems biology has led to a new regime of inquiry being formed. Transfers of tools, methods and explanations from one domain of inquiry to another are important means of producing integration, and we think the transferability of such systems of research is likely to be driving molecular systems biology into new perspectives on biological processes as well as the creation of new fields and new accounts of scientific methodology. (O'Malley and Soyer 2012: 64)

Indeed there is a strong sense in which 'the creation of new fields and new accounts of scientific methodology' is precisely the aim of cognitive science or at least its original purpose. It remains to be shown that the position of languages as systems of more abstract structures partly on the model of systems biology I have advocated, in Chapter 5, is the right approach to achieving integration and intersection. In the next section, I make the case for this claim and in the final one I make the further connection with cognitive neuroscience and structuralism.[11]

9.3.2 The Case for Language at the Intersection

The argument for using language and linguistics as an intersectional and integrational tool for cognitive science does not preclude psychology or another field from occupying a central place in the field in addition. The aim is merely to offer reasons for reconsidering the move away from language associated with the Second Generation Cognitive Science. The latter term was originally coined by Lakoff and Johnson (1999) to describe a new approach to the study of mind, one which relinquished the computational metaphor for an embodied and situated understanding in which cognition is continuous with culture and bio-evolutionary mechanisms. Cognitive linguistics was proposed in order to incorporate an opposing view of language to that of the generative tradition. On this view, any claim about language is subject to evidence from neurobiology, psychology, and cognitive anthropology.[12] Lakoff (1991) calls this the 'cognitive commitment' or the idea that one should "make one's account of human language accord with what is generally known about the mind and brain from disciplines other than linguistics" (54). This is certainly a sound principle. However, if we consider the history of linguistics within the classical cognitive scientific tradition we can appreciate adherence to this principle in many forms. In addition, there seems to be a reverse order of explanation within cognitive linguistics, suggested by Lakoff above, which threatens the place of language as a core cognitive phenomenon.

It is not a novel claim to state that generative grammar has had a number of interdisciplinary influences before its official inception and during its existence. The more interesting claim is related to how it can be argued that mentalism was the aspect that connected it to these various disciplines, many of which form the core and periphery of cognitive science. This is precisely my claim. Thus, I hope to show below that the methodology of generative linguistics is informed by the methodologies of mathematical logic, computer science, philosophy, psychology and more recently biology. A fact that advocates strongly for its central place within cognitive science. Chomsky has repeatedly emphasised that language *qua* natural object should be studied both with the resources of empirical science and towards some form of unification with other sciences:

> I would like to discuss an approach to the mind that considers language and similar phenomena to be elements of the natural world, to be studied by the ordinary methods of empirical inquiry [...] A naturalistic approach to linguistic and mental aspects of the world seeks to construct intelligible explanatory theories, taking as "real" what we are led to posit in this quest, and hoping for eventual unification with the "core" natural sciences. (Chomsky 2000a: 106)

The history of generative grammar is a broad and complex topic. It is not my purpose to delve into that here (see Newmeyer 1996, Tomalin 2006, Lobina 2017 for some notable attempts). Rather I hope to show that many of the initial and more recent interdisciplinary inspirations were sought for precisely the purpose of understanding the mind in the naturalistic and unified sense above.

The first of these insights was adapted for the purpose of explaining linguistic creativity or a related formal property of discrete infinity. Mathematical logic, especially formal proof-theory, offered linguists a way of capturing a universal linguistic property which Chomsky (2000a) suggests seemed like a contradiction before such developments, naming how a finite mind can encompass an infinite capacity for expression production and comprehension.

Notice that on a Platonist understanding of linguistics, there would be no obvious cardinality issues since languages could comfortably be considered discretely or even nondiscretely infinite.[13] However, if linguistics is to be a tool for understanding the mind, then an explanation is called for as to how an infinite capacity can be modelled on a finite instrument. The adaptation of the mathematical concept of recursion from logic and early computer science was especially harnessed in linguistics since the time of early syntactic theory. The work of Hilbert, Turing, Post, Kleene, Soare, and others provided the basis for understanding recursive and computational systems generally. As Lobina (2017: 38) notes,

> [I]t is important to point out that the results achieved by mathematical logic in the 1930s provided a mechanical explanation for a property of language that had been noticed a long time before; namely, the possibility of producing and understanding an incredibly large, perhaps infinite, number of sentences.

Thus, recursive, generative grammars defined in terms of finite rule systems became a mainstay of formal linguistics. A neglected aspect of this methodology is that part of its motivation was the explanation of a linguistic property *qua* mental state or cognitive structure.[14]

The next interdisciplinary influence to mention was naturally from analytic philosophy. Due to the Linguistic Turn, the influence of Wittgenstein and formal logic, 'first philosophy' was mostly considered to be the philosophy of language. A naturalistic understanding of the latter became essential to banishing the recalcitrant debates surrounding mind-body dualism and the emergence of consciousness. Quine's approach to these issues was naturalistic but behaviouristic and thus parochial in scope. Advances in the philosophy of mind such as identity theory and the aforementioned LOT hypothesis provided additional support for a mentalist and internalist understanding of language and even semantics (traditionally associated with issues of reference and correspondence). Again, the engagement with philosophy (although antagonistic at times) proved to be centred around the mentalism of linguistics and its possible extension to other areas of the philosophy of language such as semantics.[15]

In terms of the historical role of philosophy, Chomsky's programme benefited greatly from the work of Nelson Goodman on constructional systems theory. But "Chomsky borrowed more from Goodman than merely some of his ideas concerning simplicity measures in constructional systems" (Tomalin 2006: 121). His 1953 paper, 'Systems of syntactic analysis', aimed at connecting concepts from his mentor Harris' linguistic analysis to the work done by Quine and Goodman. Carnap's work on logical syntax also provided useful tools for linguistics especially concerning the development of a formal, in the sense of nonsemantic, computational system that relates to natural language. His influence was already felt indirectly in the philosophy of Quine and Goodman from which linguists gained insights and in a more direct way, his work informed Bar-Hillel's first recursive treatment of syntactic structures.

On the negative side, the aforementioned history also inspired one of generative grammars most controversial claims, namely *the autonomy of syntax*. We have encountered this claim, which basically amounts to an isolation of syntax as a separate purely linguistic computational system which interfaces with semantics and phonology as epiphenomenal features. Jackendoff (2002) goes as far as to claim that this posit, what he calls 'syntacto-centricism', is essentially a 'scientific mistake' and designs his own theory, the Parallel Architecture, in direct contrast to it. Cognitive linguistics too eschews the autonomy of syntax posit in favour of a more integrated approach in which semantics, pragmatics and general cognitive function can govern syntactic phenomena.

The relationship between psychology and linguistics is perhaps best exemplified by the so-called *derivational theory of complexity* (Miller and Chomsky 1963) which ultimately led to the development of psycholinguistics, an important albeit peripheral member of the cognitive scientific programme. The idea was that the linguistic structures posited by formal theory would find realisation in psychological processes and not only this but the effects could be measured. A version of the claim is that "the complexity of a sentence is measured by the number of grammatical rules employed in its derivation" (Fodor et al. 1974: 319). However, this project was largely unsuccessful and yielded no real connection or significant empirical results. The competence and performance distinction proved more problematic than initially thought. Jackendoff (2002) describes the distinction in terms of 'hard' and 'soft' idealisations. He claims that a soft idealisation is a matter of convenience with the ultimate aim of a reintegration of excluded material. The competence-performance distinction based on this latter concept would be a more practical division of labour. 'Harder' idealisations are self-contained and severed from the excluded factors in such a way that it is not possible to reintegrate them. He likens the situation in linguistics to that of the distinction hardening over time.

Nevertheless, the failure of the posits such as the derivational theory and the general quest to find formal rules in psychological processes eventually led to a new field being born. Psycholinguistics uses its own methodology and theories to investigate real-time linguistic processing. Although there are a number of interesting more recent proposals linking formal linguistics to real-time processing such as Lobina (2017), Martorell (2018)—especially to theories under the Minimalist Program (Chomsky 1995)—Christiansen and Chater (2016) offer an alternative multilevel representational account of language based on incremental parsing considerations and the 'Now-or-never' bottleneck or the fleeting nature of linguistic input, i.e., "[i]f the input is not processed immediately, new information will quickly overwrite it" (Christiansen and Chater, 2016: 2). Thus, they advocate that processing constraints be incorporated into the grammar and formal linguistic models themselves essentially collapsing the distinction between competence and performance. Dynamic syntax too incorporates processing within the theory of grammar, rejects the competence-performance distinction, and roots its idealisations in the "dynamics of real-time language activities" (Cann et al. 2012, p. 359) where semantic factors inevitably affect any analysis. In addition, it eschews the sentential string-level idealisations of FLT and many of the approaches we have been discussing in the book.

Another issue that connects generative linguistics to psychology is in the form of the data it uses to construct its theories. Introspective judgements and the judgements of native speakers on linguistic examples form the main source of data for theory construction. However, the reliance on intuitions have also resulted in a number of criticisms, which we cannot delve into here.[16] Psychology itself has been less receptive to linguistic theory for numerous reasons; its formalism, data collection techniques and, of course, the many theory changes within generative linguistics did not help matters either: "[w]hen Chomsky made significant changes to his theory of grammar, this discouraged many psychologists" (Chipman 2017: 5). The work done in Chapter 5 and the structural realist framework of this book should go some way in alleviating those worries.

The last discipline which has had an impact on generative linguistics also has tentative membership in the cognitive science club, the burgeoning field of evolutionary biology. The inclusion of this discipline has been challenged by Fodor and Piatelli-Palmarini (2010) as having little to no value for cognitive science. Nevertheless, the linguistic enterprise, under generative grammar, has redefined itself as a subfield of biology, so-called biolinguistics (Lenneberg 1967) in which evolutionary concerns are paramount. There are also cognitive scientists who similarly claim that the field is continuous with biology (Pinker 1997, Carruthers 2006). The study of Universal Grammar itself has sometimes been claimed to take its cue from biology.

UG has been studied in a modular fashion, as a mental organ, a coherent whole that can be studied in relative isolation of other domains of the mind/brain. Indeed studies of various pathological cases have shown this mental organ to be biologically isolatable. (Boeckx 2005: 46)

And even if biology *simpliciter* seems to be a controversial bedfellow to the other cognitive sciences, contemporary evolutionary biology seems directly relevant to any overarching research agenda. Many studies in language evolution take the view that language and cognition co-evolved (e.g., Bickerton 2014a). This possibility alone, if it were shown to be true, would make a strong additional case for why linguistics and language should occupy a central role in cognitive science.

Lastly, psychology does not embody a similar relationship with other cognitive sciences in terms of history or methodology. Thus, on an intersectional approach, psychology would fare somewhat less favourably as compared with linguistics.

At this junction, an objection might be raised. Perhaps generative linguistics was initially a pioneer in cognitive science and had interdisciplinary roots (or ideals), but so what? Since then, linguistics has gone on to be isolationist in methodology and disconnected from other fields studying the mind. This is a powerful objection. Historical intentions are hard to determine and when they can be fixed, they do not ensure present day sentiment.

This is where the philosophical framework developed in the book so far comes into play. For instance, adopting a multiple formal models approach as suggested in Chapter 5 provides a better candidate understanding of linguistics as a whole since it would include insights from other grammar formalisms and thus cover more empirical ground. This understanding in turns provides a strong claim for integration with the cognitive sciences as it rejects various aspects of CTM and is informed by processing and cognitive constraints (for instance, in the case of dependency grammar and construction grammar respectively). Similarly, the biological connection assumed in generative linguistics is better pursued in terms of systems biology as argued in Chapter 5. This view, as we have seen, allows us to integrate aspects of the linguistic community and environment into our models thereby connecting with 4E approaches to cognition. In other words, the picture of linguistics I have been motivating in terms of structuralism and structural realism supplements many of the aspects of the generative tradition which proved recalcitrant to the paradigm shift in cognitive science.

Nevertheless, admittedly the argument so far has merely indicated that cognition or cognitive structure was at the heart of many of the interdisciplinary methodological choices of the generative paradigm in linguistics. The last argument I will make suggests that the structural realist interpretation of linguistics advocated in the book so far can serve as an ontological connecting point to the cognitive sciences more generally.

9.4 Unifying Cognitive Structures

The previous sections offered a structural, albeit architectural, argument for an intersectional explanatory approach to the cognitive sciences. Using linguistics as an intersectional science proved valuable during the classical cognitive revolution. I have argued that if fruitful collaborative endeavour is the goal, then the methodology offered by the structural study of language is pivotal.

Therefore there is a more ontological reason to favour an intersectional approach to the cognitive sciences, namely that by doing so we can hope to illuminate cognitive structures themselves. Here, the kind of structural realism advocated in Chapters 4 and 7 becomes relevant once more. In cognitive neuroscience, graph theory is generally used to represent brain networks. And brain networks come in two overarching types. There are anatomical or structural networks identified by means of Magnetic Resonance Imaging (MRI) techniques such as diffusion tensor imaging (DTI) in which the diffusion of water molecules are used to study neural tracts and the white matter organisation of the brain. Then there are functional networks which are not completely reducible to anatomical connections and thus not completely amenable to the latter MRI (and fMRI) techniques. They are composed of "patterns of statistical dependence among neural elements" or functional connections (Sporns, 2013: 248). Yan and Hricko (2017: 2) describe the difference in the follow way:

> For example, if activity in one brain region occurs when some other brain region is active, and vice versa, there is a functional connection between those two regions, in which case they exhibit functional coupling. Functional connectivity is not nearly as stable as anatomical connectivity, and can change in the course of tens of milliseconds.

Nevertheless, in both cases graph theory is an essential tool in mapping and identifying brain networks. Both kinds of networks are represented by means of undirected graphs in which nodes serve as brain regions and edges as connections, anatomical or functional depending on the context. When neuroscientists model brain networks, the models they use are graph-theoretic structures similar to the case of grammars discussed in Chapter 4. Most linguists use undirected, connected, acyclic graphs since that is what a tree is (as we saw in Chapter 7). Thus cognitive neuroscientists and formal linguists tend to use similar mathematical structures in analysing their respective data sets. Of course, many theorists across many disciplines use graphs in their data analysis so the connection might not be more than a coincidence or convenience (or both). But there is a sense in which both neuroscientists and linguists assume structural characteristics of their targets based on the graph-theoretic models employed.

To be more precise, I will show that a *common argument pattern* can be discerned across disciplines in terms of identifying structure. The common pattern in this case is the application of multiple graph-theoretic models in search of invariance across them. Here I follow Kitcher's (1989) account of unifying scientific explanation in which maximisation of the ratio between the explanation of the phenomena is inversely proportional to the causal processes evoked. As Ladyman and Ross (2007: 31) put it, "[a]n argument pattern is a kind of template for generating explanations of new phenomena on the basis of structural similarity between causal networks that produced them and causal networks that produced other, already explained phenomena". In fact, they go on to define scientific knowledge as a strategy of problem-solving involving the communication of such patterns viewed as structures.

One reason to suspect that there is something more structurally significant afoot in cognitive neuroscience is related to a problem we've encountered on a number of occasions so far. In Chapter 7, I called it 'the problem of multiple grammars' and related it to Benacerraf's famous problem of multiple reductions. Recall that the latter was meant to showcase the superiority of a structuralist ontology over certain rivals, particularly platonism. There is a similar issue within cognitive neuroscience. The issue starts with a necessary precondition for the application of the brain imaging techniques mentioned above, namely *parcellation*. Parcellation is the segmentation of the brain into regions in accordance with the nodes of networks. Sporns (2013: 249) describes two ways of achieving these divisions:

[B]y parcellating cortical and subcortical gray matter regions according to anatomical borders or landmarks, or by defining random parcellation into evenly spaced and sized voxel clusters.

Once this task is completed, a number of graph theoretic tools are at the disposal of theorists, including path lengths between connections, integration based on interconnections within the networks etc. One related notion is that of a *hub* or a node which has the most number of edges attached to it. In Chapter 7.4, we saw dependency graphs which identify root nodes to which other words are attached. A root is then like a hub in a structural sense. Getting back to the problem at hand, given that there are different ways to parcellate the brain or 'parcellation schemes', the question of which is the most accurate scheme becomes important. Do the nodes chosen in a particular parcellation scheme map onto real items in brain networks? Yan and Hricko (2017: 4) describe the complete situation in the following manner:

> The challenge that they face here is not restricted to the nodes. Since edges are investigated only once nodes are well-defined, the edges and the resulting graphs are also parcellation-relative. In contrast, the brain networks that cognitive neuroscientists seek to investigate are presumed to exist independently of the choice of parcellation scheme—a real network must be parcellation-independent.

Thus, the graphs need to directly or accurately represent real networks in the brain. In order to do so, the real content of the theories need to be identified among the many possible groupings. Of course, to a philosopher of science, this scenario is quite commonplace and usually goes under the banner of 'underdetermination of theory by evidence'. French (2014) claims that dealing with this issue is one motivation behind the move from realist to structural realist thinking.

To see the structurally relevant aspect of the parcellation problem, one has to appreciate how it is often resolved in neuroscience. Basically, neuroscientists construct multiple distinct graphs and look for invariant structure across them. We've seen and advocated for this strategy in Chapter 7 for linguistic grammars. As Sporns (2014: 653) notes "studies of brain networks using a variety of parcellations [...] have converged on a set of fundamental attributes of human brain organization that are largely consistent with those found in nonhuman primates". These studies have uncovered empirically significant features such as robust hubs in particular brain regions. In some rare cases, neuroscientists can confirm such findings by means of tools such as ablation in which damaged parts of the brain are removed (see Harat et al. 2008 for a history of this surgical tool). Nevertheless, the strategy of employing multiple graphs and checking for convergence has proven successful in many cases. Essentially, this is a practical application of the concept of *intersection* discussed in the previous sections. By applying multiple intersecting mathematical models to a given cognitive target, in search of invariant structures, we can partially shed light on the cognitive structures underlying of mental competence, i.e., the real cognitive patterns.

The structural analogy is so strong that Yan and Hricko (2017) use the above characterisations to suggest a localised version of structural realism for cognitive neuroscience. In other words, the models of the field furnish us with structural knowledge of functional and anatomical brain networks. I have argued in earlier chapters that our knowledge of grammatical and linguistic reality is similarly structural.

Beni (2019a, 2019b) makes an adjacent structural realist case for computational neuroscience and the unifying nature of the 'Free Energy Principle' (FEP) which brings together variational Bayesian brain theory, work in neuroplasticity, evolutionary theory, probability theory and a number of other fields. FEP models the mechanism of the brain's tendency towards minimising entropy (viewed as discrepancy between internal models and external causal structures) as a Bayesian measure of uncertainty. He states that "[i]t is in virtue of its inclusive mathematical formalism that FEP could relate insights from these [aforementioned] diverse fields" (Beni 2019: 154). Intersectional mathematical structure again breeds insight into cognition.

Whereas Yan and Hricko argue that the appeal of structural realism is local, Beni aims toward global explanation and unification in cognitive science. Again, he too follows Kitcher's (1989) definition of unification as the reduction of the number of facts we ultimately have to accept by recognising recurrent patterns of derivation. Citing the unifying role of ontic structural realism in physics, Beni claims that:

> [T]he state of metaphysical underdetermination in physics bears resemblance to cases of explanatory pluralism in cognitive sciences. Accordingly, it would be possible to face the challenge of pluralism in the field of cognitive sciences by a structural realist strategy. (2019a: 162)

He specifically argues against the more pluralist (or multiple models) approach of critics of FEP such as Colombo and Wright (2021). The latter authors adopt a mechanistic understanding of the field and thus argue against the unificatory appeal of variational Bayesianism and principles such as FEP. But Beni counters that by appreciating a structural realist interpretation of the aforementioned fields pitched at a metatheoretical level, the possibility of uniting disparate elements and disciplines within the cognitive sciences is activated. I make a similar claim here for the models of formal linguistics and their graph-theoretic instantiations. Again, the Bayesian framework is claimed to provide us with structural explanations of cognition (and its interaction with the environment).[17]

If cognitive structures are susceptible to the intersectional approach on the microscale, then the cognitive sciences should be amenable to a similar approach on the metascientific level as is the case with the FEP. The possibility and success of aiming various structural analyses, including formal grammars, towards a common cognitive problem further emphasises the relevance of methodological architecture in studying the mind.

Common argument structure is one way of tracking real patterns. If such a connection can be established between the view of linguistics I have presented in the book and aspects of cognitive neuroscience, then we've gone some way in showing that integration is indeed possible. Linguistically real patterns and cognitively real ones might have significant overlap. This was the somewhat unfulfilled promise of generative grammar. But with the tools of structural realism and the concept of real pattern analysis, we can make a stronger case for this possibility here.

There is, of course, much more work that needs to be done. This chapter was in part a path clearing mission. Now we should be in a better position to see the trees and distinguish them from alien vegetation.

10

Conclusion

A Canopy in the Rainforest

It's been a long journey into the philosophy of linguistics and in its course, I have marshalled both insights and theories from naturalistic philosophy, structural realism, and the general philosophy of science to produce an ambitious new look at the field of linguistics from a philosophical point of view. Throughout, I have more than hinted that there is something especially special about this science given its formal underpinnings and naturalistic aspirations.

There have been some notable volumes on the philosophy of linguistics, both in the past and recently. Ludlow (2011), Pereplyotchik (2017), and Rey (2020) are recent offerings. What these books share with mine is an appreciation of the fascinating philosophical depth of theoretical linguistics. Where we diverge is in focus. I have chosen to be considerably more ecumenical while they have worked closely within generative linguistics. I have also chosen to mine the resources of the philosophy of science for insights more directly and comprehensively. The result has been a work that covers a frightening amount of theoretical options from model-theoretic syntax to cognitive linguistics to formal semantics all the while claiming some family resemblance within a broadly structural realist perspective.

In this short concluding chapter, I want to trim some dangling branches and suggest conduits for further research.

The broad research programme into which my project falls is what Ross (2000) calls 'Rainforest Realism' in that it's a type of realism that is 'lush and leafy'. It is an ontological view derived from Dennett's real pattern analysis and opposed to the Quinian 'value of a variable' doctrine. It espouses the reality of patterns far and wide while imposing only two conditions on their existence, (1) that they are projectible from at least one physical perspective, and (2) they encode information about at least one structure of events or entities in such a way that that the encoding is more efficient than the bit-map encoding of S. Both conditions block a certain kind of anthropocentricity characteristic of *a priori* metaphysics, while the second also filters the patterns *via* a kind of

Language, Science, and Structure. Ryan M. Nefdt, Oxford University Press.
© Oxford University Press 2023.
DOI: 10.1093/oso/9780197653098.003.00010

the non-redundancy test. I modelled my own definition of linguistically real patterns on a modified version of this idea.

Rainforest realism then equates existence with measurement and allows for measurement to take various forms from various special sciences (with due homage paid to physics in the form of the PPC and PNC). I have here attempted to carve out a canopy in the rainforest where linguistics is a special science which investigates one of the most fundamental aspects of human existence, our ability to produce, acquire and understand natural languages. These natural languages are in turn special kinds of patterns with their own existence conditions. Unlike many other special sciences, linguistics also has highly developed formal measurement techniques, some of which have natural translations into the information-theoretic frameworks described by structural realists like Ladyman and Ross, Floridi, and others.

One not-so-subtle aspect of many of the views under discussion in the book can potentially be viewed as closet computationalism. Perhaps I can be justly accused of assuming some sort of computationalist perspective and discreteness of the target phenomena in many places. My view, however, is that computationalism is not incompatible with naturalism or even more contemporary non-representationalist views in cognitive science as I have shown in Chapters 4, 5, and 9 with the marriage of FLT and biological systems, and generative grammar with cognitive linguistics and 4E approaches to cognition. The rainforest is rich in flora.

Hutchins (1995) produces a fascinating case study of computational cognition onboard a naval vessel undergoing complex navigation during an emergency procedure. There is no central controller, only a distributed team of individuals all playing essential roles without which catastrophe would ensue. Navigation, a computational property, emerges. His view of 'natural situated cognition' invites theorists out of the laboratory to witness distributed cognition 'in the wild'. But the view shows that computationalism is not only applicable at the individual level and, in fact, might be more appropriate at the social coordination level. Therefore, social cognition can also be computational if placed within a complex systems understanding of the community or group in question. In Chapter 5, I made similar moves to place language within the complex biological system of the redefined linguistic community, beyond individuals and even human interlocutors. In both cases, the environment, with all of its dynamics, is of paramount importance for cognition and language, respectively.

The full account of the present work can best be represented by the central insights of each chapter, which I present below. The significance I hope to have achieved mostly concerns the twin goals of promoting the philosophical richness of the special science of linguistics, especially from a philosophy of science perspective, and elucidating the concept of linguistic structure so seminal to the enterprise since its very formalist beginnings in the early twentieth century to the present day.

The thematic clustering of the constituent parts of the overall view are as follows. The first two insights are path-clearing in nature. They argue for a reconceptualisation of the older debate in the ontology of language and a rejection of object-oriented accounts which have predominated that sometimes incendiary discussion.

CENTRAL INSIGHT I: OBJECT-ORIENTED ACCOUNTS OF THE ONTOLOGY OF LANGUAGE ARE NONSTARTERS AND ANY FOUNDATIONAL VIEW WHICH ENDORSES THEM IS THEREFORE LACKING.

CENTRAL INSIGHT II: ALTHOUGH ANTIREALISM AND PLURALISM BOTH POINT TO IMPORTANT EXPLANANDA IN THE ONTOLOGY OF LANGUAGE, THEY FAIL TO PROVIDE ADEQUATE ACCOUNTS FOR ADDRESSING THEM.

The third and fourth insights proffer a novel naturalistic ontology in terms of real patterns and the linguistic systems, modelled on biological systems, which ground those patterns. The fifth insight merely extends the view to a case study in terms of the ontology of SLEs and words specifically.

CENTRAL INSIGHT III: LANGUAGES ARE NON-REDUNDANT REAL PATTERNS, THE STRUCTURES OF WHICH ARE CAPTURED BY FORMAL GRAMMARS QUA COMPRESSION MODELS.

CENTRAL INSIGHT IV: NATURAL LANGUAGES ARE EMERGENT PHENOMENA WITHIN DYNAMIC COMPLEX BIOLOGICAL SYSTEMS COMPRISING NETWORKS OF INTERNAL MECHANISMS, EXTERNAL CONVENTIONS, AND ENVIRONMENTAL FACTORS.

CENTRAL INSIGHT V: WORDS, PHRASES, AND RULES ARE ON A STRUCTURAL CONTINUUM WHERE THE ROLES THEY PLAY IN OVERARCHING LINGUISTIC STRUCTURES SERVE AS THEIR PRIMARY ONTOLOGICAL STATUS.

From precisifying the naturalism often touted in the philosophy of linguistics and purely ontological concerns, the sixth and seventh insights are more concerned with the philosophy of science. Insight VI takes on the history of generative linguistics and provides an anti-pessimistic account of theory change therein. It also goes further to use this analysis as a connecting point to other non-generative theories of grammar all the while putting forward an ontic structural realist view of the science itself. Insight VII looks beyond syntax to semantics and the nature of the interfaces between different linguistic systems. Here informational structural realism and computer science inform the discussion of levels of abstraction and their relations in linguistic theory. Finally, a new position in metasemantics is advanced (in sketch form) along structural lines allowing for a more complete account of the philosophy of linguistic theory.

CENTRAL INSIGHT VI: THE PRIMARY VEHICLE OF THEORY CHANGE WITHIN GENERATIVE LINGUISTICS AND THEORY COMPARISON ACROSS FRAMEWORKS IS STRUCTURE.

CENTRAL INSIGHT VII: NATURAL LANGUAGE IS A COMPLEX SYSTEM CONTAINING LRPs ANALYSABLE AT DIFFERENT 'LEVELS OF ABSTRACTION', EACH CONNECTED BY A NESTED 'GRADIENT OF ABSTRACTION'.

In the last chapter, and its final insight, the aim is to take all the structural thinking harnessed and set forth in the earlier parts of the book towards a burning question in the philosophies of linguistics and the cognitive sciences, namely how is the scientific study of language connected to the scientific study of mind? Here the view draws from the history and motivations of the place of linguistics in the classical cognitive revolution and makes a case for the continued reflection on language in the future cognitive sciences utilising structural realism as a conduit. It's structures all the way down the canopy!

CENTRAL INSIGHT VIII: LINGUISTIC THEORY, VIEWED STRUCTURALLY, DOVETAILS WITH SOME OF THE MODELLING PRACTICES OF THE COGNITIVE SCIENCES CONSTRUED IN TERMS OF STRUCTURAL REALISM.

Despite the many vicissitudes of this particular philosophical journey, there are still many avenues to travel, many structures to uncover and catalogue. Unfortunately, that extended task is best left for another occasion. My hope is that those scholars interested in linguistic structure as well as those interested in structure in general will draw some inspiration from the account provided in this book and that it might spark further reflection on the topic going forward.

Notes

Chapter 1

1. See Katz (1971) for a book length defence of such a view. Of course, Katz' position contrasts sharply with that of the ordinary language philosophers in obviating surface linguistic form and mining philosophical truth at a deeper level brought out by the generative grammar of the time. He interestingly refers to this concept as the "second linguistic turn".

2. In the SEP article on the topic, Scholz, Pullum, Pelletier, and Nefdt (2022) take a similar line on the distinction between the philosophy of language and the philosophy of linguistics.

3. Here I follow Blutner (2011), although I have substituted his third tenet for the Universal Grammar postulate and my description of the rule-based view is different to his.

4. Kuipers (2007) distinctions are designed for the sciences generally but usefully adapted by Tomalin (2010) for the development of generative linguistics. In this framework, there is a hierarchy of scientific categories roughly as follows. At the top are (1) research traditions, e.g., Generative Linguistics itself (including phonology, syntax, etc.), which are instantiated by (2) research programmes such as generative grammar (further subdivided into Standard and Extended Standard Theory, Minimalism, etc.) or the parallel architecture, optimality theory which in turn have (3) core theories (such as syntactocentricism or recursion) and finally (4) specific theories of particular phenomena which share core theoretical tenets. "This seems reasonable since the phrase 'generative grammar' is standardly used to refer to different theories of generative syntax that have been proposed during the period 1950s–present, and, given this, it would be misleading to classify GG as being simply a 'theory'" (Tomalin 2010: 317).

5. There is a distinction between the idea that there are universal typological properties which are common to all natural languages (in surface syntax, morphology, etc.) (for instance, the wealth of research which flowed from Greenberg (1963)), and the claim that there are underlying universal principles of human language which are relevant to explaining phenomena such as language acquisition. GG is only committed to the latter. Chomsky (2005) also indicates how 'third factors', such as those in the P&P model, can significantly constrain UG. I thank Gabe Dupre for pointing this out to me.

6. See Dupre (forthcoming) for a discussion of defective data with specific reference to the competence-performance distinction in generative linguistics.

7. Technically, 'rule-based' and 'representationalism' are two separate claims. But they do seem contingently inseparable within the field.

8. See Joseph (2015) for more on the arbitrariness of the sign and Hjelmslev's (1943) on a particularly radical structural reconstruction of Saussure's views which informed much of formal functionalist approaches to linguistics.

9. It is also argued by some, that structuralism emerged in response to Benacerraf's (1965) 'multiple reductions' problem exemplified by the incompatibility of two equally legitimate set-theoretic formulations of arithmetic in terms of the von Neumann ordinals and the Zermelo numerals. I see a deep analogy with this issue and the 'multiple grammar formalisms' problem I identify in Chapter 7.

10. Of course, things are rarely this simple. For instance, one could reject instrumentalism and still, following Kuhn, emphasise the theory-laden nature of observation. Kuhnian skepticism about science would then amount to a view that we have no chance of extending our understanding beyond the paradigm we are currently in, i.e., we are constrained by the limitations of our present science. I thank Harold Kincaid for pressing me on this alternative.

11. Which as Frankish and Ramsey (2012) state, in their handbook on cognitive science, is sometimes indistinguishable from cognitive science itself.

Chapter 2

1. One might also add scepticism of meaning to the list. Interestingly, meaning-scepticism persisted long into the generative movement and in part resulted in the so-called 'Linguistics Wars'. See Newmeyer (1996) for discussion.

2. The first traces back to Frege's seminal work in the *Begriffsschrift* (1879) and *Grundgesetze der Arithmetik* (1893) on the formalisation of arithmetic which ushered in the use of predicate logic in the philosophy of language. The second has a history dating back to the Hilbert programme in metamathematics and the introduction of proof-theoretic techniques in formal syntax through the work of Carnap and others. The former notion consisting of the familiar functions, individual constants, variables, and predicates derives from the idea of a deductive system in logic. Basically, a deductive system comprises a set of axioms and rules of inference that can be used to derive theorems of that system. The system preserves truth in accordance with valid applications of the rules. The latter notion involves simply taking languages to be (possibly infinite) sets of strings over a finite alphabet, i.e., a calculus. These languages can be characterised by devices known as *grammars*, which essentially specify the well-formed formulas of the language (it tells us which strings are in the language and which are not).

3. Kac (1992) provides an interesting approach in this regard. He argues that the subject matter of linguistic theory is grammaticality, following Chomsky (1957), not the 'language faculty' or competence as later generative accounts assume. He further aims to resurrect a notion of traditional grammar presupposed in most contemporary linguistics through his 'etiological analysis'.

4. For Katz (1996) the abstractness concern in linguistics is a special case of the general problem of abstractness in the formal sciences. An account such as the strict finitism or 'inscriptionalist nominalism' characterised by the Hilbert programme, for instance, failed as an appropriate interpretation of mathematics according to Katz. In order to capture the infinity of mathematics via the empiricist scruples of nominalism,

only reconstructed language about the infinite is permitted, "mathematics is about mathematical expressions" (Katz 1996: 273). The objection is simply that to make sense of such talk, we need either expression types, which take us back to abstract objects, or expression tokens, which need to allow for unactualised *possibilia* which in turn are no less metaphysically suspect than abstract objects. Katz, however, neglected the vast literature on actualist reinterpretations of quantified modal logic, some varieties of which posit *contingently nonconcrete* objects in an attempt to avoid commitment to *possibilia*.

5. Although it must be noted that Devitt distances himself from the standard Chomskyan criticism of an E-language conceived of as wholly disconnected from a theory of linguistic competence. Devitt's account involves linguistic outputs as the outputs of mental competence. But to the extent that his view deals with *externalia*, I have kept the term.

6. There are a few reasons usually offered in favour of the general theoretical position described above. One motivation, in terms of the philosophy of science, is that the most prominent alternative, that of studying an E-language or external language, is often said to be scientifically untenable. E-language, it is claimed, are political entities determined by the vagaries of convention or fiat (see Chomsky (2000) for his famous 'London example' arguing that there is no clear external object to which 'London' refers). It is argued that there are no clear boundaries between external languages, their features are unstable and therefore their study is outside the purview of serious science. However, as Santana puts it "[a]ll science idealizes, so to reach the punchline of his argument Chomsky needs to further show why the idealizations leading to a technical notion of E-language are illicit" (2016: 510). This, however, is precisely the claim, i.e., that there are no law-like generalisations to be found with E-languages. I'm not sure this claim is sustainable but either way it is orthogonal to the issue at hand.

7. However the issue of normativity in linguistics is an open one. Ludlow (2013) makes a case for a normative (not prescriptive) interpretation of generative linguistic rules. Similarly, Kac (1994) and Pullum (2007) also provide illuminating discussions of formal linguistics and normativity.

Chapter 3

1. This view resembles modelling accounts of linguistic theory in some ways in which theoretical posits are indirect representations of target systems. In this way, we are not committed to the existence of the posits of the models. See my (2016, 2019a).

2. The idea of linguistics proper has antecedents in Hockett's (1958) distinction between what he calls central and peripheral linguistic subsystems.

3. I abstract over the necessity claim of the social aspects of natural language which involves phatics and other similar speech acts. See Millikan (2005) and Peregrin (2015) for two distinct approaches to the legitimacy of public language as a object of scientific inquiry.

4. These are technically obstruents which create small pockets of air and then release them in the production of loud consonants. isiXhosa and isiZulu are click languages of this sort.

5. Of course, a different scientist could aim to abstract away from semantics by the same procedure. In fact, the arguments for 'autonomy of syntax' often presented themselves under this guise. See Sampson (2001) for an argument for the non-scientific/non-empirical nature of semantics based on similar reasoning.

6. Unless we say much more about the connection between the mental and the mathematical in terms of either intuitionism or finitism etc.

Chapter 4

1. Kincaid (2013) looks across a range of naturalistic or scientific philosophical accounts to identify two core tenets present in all of them: "(1) an extreme scepticism about metaphysics when it is based on conceptual analysis tested against intuition [...] and (2) the belief that scientific results and scientific methods can be successfully applied to some problems that could be called metaphysical" (3).

2. You might worry that this position erroneously dodges the 'Great Method Debate' in the history and philosophy of science. It surely does. But I am not certain that delving into deep unsolved (unsolvable?) issues on the nature of the scientific method is a useful exercise. On the other hand, using purely institutional/funding practices can lead to worrying situations. A parapsychological grant proposal on the effects of ectoplasm might have received funding at a serious institute in the mid-twentieth century. That doesn't make it science. See Strevens (2020) for an interesting take on the methodology and nature of science as an essentially irrational evidence-obsessed pursuit quite alien to human intellectual endeavour.

3. Some, such as Hintikka (1999), have strongly objected to the reliance on speaker and especially linguist intuitions as the primary tool of theory-building. The idea is often associated with a sort of Cartesianism about linguistic knowledge or privileged access to an internal state of knowledge, explicitly inaccessible. Linguistic intuitions, on this view, reflect the internal generative grammar or I-language of individual language users and can therefore be incorporated as data for linguists. If this is the case, then linguists themselves can provide the data needed for their theories.

   It is unclear if linguistic theory is committed to any such view. For instance, Marantz (2005) argues that the role of introspective data has been mischaracterised and the judgement of linguists' stand only as "proxies" or metadata aimed at representing and not reporting the data. So, the judgements of linguists' merely indicate the need for further corpora-based or distributional investigation for later confirmation.

4. Not to be confused with Rey's PT we saw in Chapter 3.

5. One might also worry that PPC taken as a historical principle would have resulted in the rejection of Darwin's theory of evolution, out of step as it was with the physics of the time.

6. A UTM is a Turing machine that can simulate any specific Turing machine on arbitrary input thereby fixing the issue with standard Turing machines that a different one must be constructed for every new computation to be performed, for every single input-output pair.

7. Compare this to the description of a category in category theory. "A category is *anything* satisfying the axioms. The objects need not have 'elements', nor need the

morphisms be 'functions' [...] we do not really care what non-categorical properties the objects and morphisms of a given category may have (Awodey 1996: 213).

8. The other versions of structuralism offer similar accounts. They differ, however, in important respects. For instance, the question of whether or not structures can themselves be considered mathematical objects. For set-theoretic structuralists, inspired by model theory, the answer is yes. Structures are set-theoretic entities themselves. For modal structuralists, structures are not objects of study. Hellman (1989) utilises this framework to avoid reference to individual mathematical objects all together (by replacing such talk with talk of *possible* mathematical objects or number-systems in his case), it is thoroughly eliminative. The point is that there is no one answer to the question of the nature of structures themselves, different structuralists will provide radically different accounts. Another question concerns the background logic, which varies from first-order with identity to second-order and modal logic given different accounts of structuralism.

9. With Platonism, no such manoeuvre is permitted since the two theories pick out different abstract objects and there is a fact of the matter as to which objects the natural numbers correspond, e.g., either $2 \in 4$ or $2 \notin 4$ (as Benacerraf (1965) convincingly argued). Of course, the Platonist might not be committed to there being a fact of the matter. For instance, Wright (1983) takes Benacerraf's dilemma to be an instance of indeterminacy of reference.

10. Elsewhere, I have made the case that formal linguistic grammars are scientific models (see Nefdt 2016, 2019a). I'll make a similar case below.

11. I thank Don Ross for this nice example.

12. The 'nondeterministic' part is less relevant to our current aims. In a finite state automaton, non-determinism amounts to the feature that a given state does not uniquely determine the next transition symbol as there could be more than one reachable state based on that state. Empty states ensure this property.

13. We'll deal with this situation in a different but related manner in Chapter 7.

14. For more on the foundations of formal language theory, see Hopcroft et al. (2001), Levelt (1974) for a classical treatment. See also Miller (1999) for more on strong generative capacity.

15. There are, of course, other tools we can use to get at structures. One such tool is the addition of *transducers* to our grammatical considerations.
    As previously mentioned, finite state automata define the class of regular languages. Finite state transducers define regular relations which are ordered pairs of symbols from two different alphabets. In other words, the accepting states of transducers are pairs of symbols. One obvious use of such a device is translation in which pairs are formed from words from two different languages. Our concerns are more theoretical. Specifically, transducers allow for direct comparison between stringsets. From this it's not hard to see how one could use such a device to assign similarity or difference scores based on the number of transitions needed to get from the sequence of one set of strings to the other (if we include the empty string, since not doing so would limit measurements to stringsets of exactly the same length). In fact, composition in finite state transducers can relate two languages directly as input and output of one another. In this way, they show us which structures are needed to get from the one to the other.

This is one way of capturing Millhouse's similarity metric for model comparison which he links to compressibility.

There is a fascinating, if not somewhat complex, result which illustrates the power of transducers to extract real patterns in language. Rogers and Pullum (2008) take as their target the informal grammar presented in the comprehensive Huddleston et al. (2002) *The Cambridge Grammar of the English Language* (CGEL). Importantly, they formalise the implicit structures of the CGEL in terms of relational structures where "a relational structure is just a set with certain relations defined on it" (Rogers and Pullum 2008: 3) defined model-theoretically. Transducers similarly deal in relational structures in that one element of the pair can be viewed as the target language and the other the metalanguage. The metalanguage used there is monadic second order logic or MSO interpreted on relational structures with tree properties. MSO is a second order variant of predicate logic and its theory of trees is the set of all MSO sentences satisfied by any tree structure. They conclude that the set of grammatical strings entailed by a CGEL grammar, modelled by MSO on tree structures, will be a context-free language. The trend they follow, in establishing this result, is exemplified in Rogers (1998) where it is argued among other things that we can successfully abstract fully away from the kinds of generative proof-theoretic grammar formalisms discussed in this section to express syntactic theories purely in terms of the properties of the class of structures they license thereby focusing directly on the structural properties of languages rather than on mechanisms for generating them. This is often called the model-theoretic approach to syntax (Pullum and Scholz 2001, Pullum 2013).

16. OR (iv) there is a transducer for encoding the shortest set of transitions or rules for getting from L' to L' (see note 15).

Chapter 5

1. All except standard platonism since as we've seen it fails to pass the non-redundancy test.

2. In part, due to another controversial claim about the timeline involved in the evolution of language which is usually thought to be around 80k-100k years ago.

3. In Chapter 8, I will say slightly more about what a 'complex system' is in line with this recent groundbreaking work.

4. They too deny the possibility of an E-language as a scientifically tractable concept of language for similar reasons to Chomsky.

5. See Bickerton (2014b) for a sharp criticism of these views and the general biolinguistic framework, especially with relation to language evolution.

6. The advent of Big Data has had a profound influence on the life sciences as it has on computational linguistics. The data explosion that originated in biological research compelled the development of systemic approaches to data analysis and departures from the single gene/protein frameworks of the past.

7. For instance, Goldberg (2013) also emphasises the 'network' aspect of the approach in which "[p]hrasal constructions, words, and partially filled words (aka morphemes) are related in a network in which nodes are related by inheritance links" (1). Network and inheritance hierarchies are also utilised in formal frameworks such as Word Grammar

(Hudson 1990), Head-Driven Phrase Structure Grammar (Sag et al. 2003) and Sign-based Construction Grammar (Boas and Sag 2012). She additionally claims that construction grammar explicitly rejects the kind of thinking involved in generative grammar which postulates hidden layers of syntactic analysis (such as nonterminals).

8. One way in which to combine the I- and E-language perspectives is by Devitt's 'respect constraint'.

> [A] theory of a competence must posit processing rules that respect the structure rules of the outputs. Similarly, a theory of the outputs must posit structure rules that are respected by the competence and its processing rules. (Devitt, 2006: 23)

It is a constraint, he motivates with various examples, mostly designed to distinguish between mental competence in a particular act and the output of that competence. But it could help us specify the relationship that needs to hold between different aspects of our biological systems. In other words, the system respects the structures of mental competence posited by the grammar and also networks of the interaction of outputs governed by the community of speakers. The point is best put negatively. The idea is that the system cannot exhibit patterns that are not compatible with the structures of the grammar and the grammar cannot posit structures that are not compatible with the patterns found in the linguistic environment. The problem is that both directions seem possible in practice as argued by Davidson, Ludlow, Chomsky and others (see Chapter 3).

9. A particularly bold proposal is developed by Kohn (2013) in his mind-bending anthropological study of the region and Runa people of Avilá in Ecaudaor's Upper Amazon. In this work, Koln attempts to de-anthropomorphise the field of anthropology by exploring the complex symbolic and (Piercean) semiotic interactions between humans, non-human animals, plant life, and general environment of a place where they come together in myriad ways. A pertinent construction of his is the idea of 'trans-species pidgins' which is a kind of language humans use to commune with animals and even the forest. It's rich with symbolism and symbiosis and a notion of a 'distributed self' which resembles the extended mind thesis in a number of interesting ways.

We do not, of course, have to agree with Koln that forests can *think*. Nor do we need to embrace the full extent of his ethnobiological perspective. His analysis only illustrates the extent of some of the systems biological ideas that can change our concepts of mind and language. In fact, there are good reasons to be cautious of some of this work from a complexity science point of view. Part of the motivation behind complexity science, and thus part of the reason the present analysis is not prey to Chomsky's methodological concerns, is that complex systems in systems biology need to be measurable and formally describable in order to be studied as such.

10. The proposal is, in essence, to treat linguistics like a complexity science and language like a complex system. There are at least one extant linguistic account which explicitly treats natural language as a complex system. In addition, there are many other such accounts in the history of linguistics. Hopper's (1987) emergent grammar is one such paradigm. Clark (1996), for instance, treats language as an emergent coordination system like a dance. Even GB of generative grammar (Chomsky 1981) might be considered a complex systems analysis of language. In that framework, the linguistic

system is divided into two classes of subsystems, those pertaining to the rule system and those pertaining to the principles. In the former, we find the lexicon, syntax and both interfaces (Phonetic Form and Logical Form). In the latter, bounding, government, Θ, binding, Case, and control theories.

> The system that is emerging is highly modular, in the sense that the full complexity of observed phenomena is traced to the intersection of partially independent subtheories, each with its own abstract structure. (Chomsky 1981: 135)

One aspect of the analysis is that language is broken up into many different component parts each with their own constraints and mechanisms. Although the theory still maintains that the faculty of language is autonomous from other cognitive systems, it does stratify the concept of UG to include the idea that a large portion of grammar is common to all languages. This becomes a move towards more complexity and modularity. The previously mentioned principles is the part of UG which acts like well-formedness conditions or constraints on the representations of each level of the grammar (D-structure, S-structure, PF and LF).

Of course, GB remains an internalist and isolationist approach to grammar along the lines of the competence model discussed above. Some streamlining of the many rules of transformational grammar takes place (*via* X-bar theory), focus on learnability is increased, and a number of subtheories are introduced. What is lacking is the interaction with non-linguistic aspects of the environment and fellow language users (through dialogue data or corpus research), the biological analogy as well as interdisciplinary methods. Nevertheless, it is the closest that generative grammar comes to being a complexity science.

The account that explicitly endorses a complexity science view of linguistics is Kretzschmar (2015). In it, he homes in on the common complex systems features of numerosity, feedback, and non-linearity or what he calls the 'A-curve' in the corpus data he evaluates. He states that "no linguist can afford to ignore the fact that human language is a complex system" and that furthermore "[a]ll approaches to human language must begin with speech, and all speech is embedded in the complex system" (Kretzschmar 2015: 2). It is unclear that generative linguists or Minimalists would agree with either statement and certainly not the latter. This is especially the case given that generative grammar has often relied on what they call 'negative data' or mistakes that language learners do *not* make, elements unlikely to be present in corpora. Furthermore, as Pullum (2007) notes, despite their merits, corpora often do not contain rare but possible constructions which can inform linguistic theory. One specific complex feature which Kretzschmar shows to be omnipresent in his corpus data is an emergent nonlinearity characteristic of market economies, namely the Pareto or 80/20 principle in which 80% of wealth is concentrated within 20% of the populace based on Zipf's Law.

> Perhaps the most striking evidence for speech as a complex system is the nonlinear distribution of the variants for any given linguistic feature. Linguists will recognize Zipf's Law, a frequency ranking of words in texts that always finds that rank is roughly inversely proportional to frequency. (Kretzschmar 2015: 24)

He claims that the Pareto principle shows up all over the data at various levels and that "we in language studies can and should make good practical use of the 80/20 Rule on a conceptual basis" (2015: 85). His own reflections on how to 'make good practical use' of the principle are slightly less compelling. After comparing the size of certain compendia of English Grammar (as evidence of the wrong kind of complexity?), he suggests that the appreciation of the 80/20 Rule should improve grammars by insisting that generative linguists focus on infrequent constructions given that they "only study just the top-ranked variants" (Kretzschmar 2015: 99). This claim strikes me as false. Generative grammar has always considered less frequent, sometimes even untokened, phrases and constructions. The debates concerning center-embedding and recursive phenomena were crucially about *possibilia* and not frequency. In fact, if there is criticism in this vein, it probably goes in the opposite direction. The mischaracterisation of generative linguistics as isolating frequent constructions and then making inductive generalisations based on these can be found throughout the book. But it is not my purpose to mount a critique of Kretzschmar's account, which is on the whole a huge step in the right direction.

11. There's a stronger version of the POV based on the kinds of mistakes children are said to make during acquisition. Given that they tend not to receive negative evidence, generativists claim that their mistakes reveal structural biases. See Chomsky (1986, 1988) for this argument based on famous examples like *Is the man who is tall happy?* See Pullum and Scholz (2002) for an empiricist critique of POV arguments in general and Cowie (1999) for a philosophical take on the issue.

12. This assumes that suitable evidence for such linguistic phenomena as parasitic gaps, hierarchical ordering rules, wh-islands, and so on, is available in the data, which I admit is a controverisal assumption at this stage of research. I thank Gabe Dupre for pressing this point to me.

Chapter 6

1. I presuppose here and throughout that some adequate definition that permits individuation of words in a text is available for each language; even that may not be true, at least not if the definition is supposed to be universal: see Haspelmath (2011) for an argument that there may be no universal notion of a word-level unit.

2. We could, of course, apply prescriptive scruples and legislate what counts as the 'proper' pronunciation, meaning and usage but this would take us quite far from the scientific study of language.

3. Linguists might interject here with the distinction between LEXEMES and WORD-FORMS where the former are the smallest units tied to non-compositional meaning (like phonemes) capable of being uttered in isolation to convey semantic content (unlike morphemes). So GO would be a lexeme stored in the lexicon, while *goes, went*, etc. are word forms related to this lexeme. Unfortunately, I don't think that this distinction helps much for the individuation problem as word-forms would be presumably at the level of types and their tokenisation would still require some ontological separation. Additionally, depending the criteria you use, you might consider GO and *went* to be different words.

4. It must be noted that homophones are indeed more difficult for children to acquire at a very early age but this difficulty does dissipate well before adolescence.

> From the age of about 4;0 children have the metalinguistic ability to understand that one word may have two distinct meanings (Peters and Zaidel, 1980; Backscheider and Gelman, 1995; Doherty, 2000). One might therefore expect children's difficulty with homonyms to decline quite rapidly after this age. However, Beveridge and Marsh (1991) showed that six-year-olds still have difficulties, and a study by Mazzocco (1997) suggests that children's difficulty in overriding the familiar meaning of a homonym persists until children are at least 10;0. (Doherty 2004: 204)

5. I thank Zoltan Szabó for this insight.
6. Citing the unidirectionality of such formalisms, he goes on to say "[p]aradoxically, this makes a declarative grammar potentially more amenable to being embedded in a model of language processing. In processing, the grammar is being used to build up a linguistic structure online, over some time course" (Jackendoff, 2018: 102). This situation is considered 'paradoxical' because generative grammars seem to naturally evoke procedural or algorithmic analogies one might expect to be more useful for understanding real-time language processing.
7. Constructions should be somewhat familiar to those working in the NLP community. Consider the idea of 'prefabs' expounded on by Poibeau (2017: 23):

> In this framework, syntax is not as prominent as in traditional approaches; the sentence is seen as an assemblage of "prefab units," or, put differently, an assemblage of complex sequences stored as such in the brain. The analysis is thus simpler, since, if this hypotheses is correct, the brain does not really have to take into account each individual word but has direct access to higher level units, reducing both the overall ambiguity and the complexity of the sentence-understanding process.

8. Perhaps the scale relativity of ontology could be appropriately applied here as well.
9. As Shapiro notes, "[t]he subject matter of arithmetic is the *natural-number structure*, the pattern common to any system of objects that has a distinguished initial object and a successor relation that satisfies the induction principle. Roughly speaking, the essence of a natural number is the relations it has with other natural numbers" (1997: 5).
10. Kaplan (2011) almost endorses a view similar to this one but insists on driving an ontological wedge between words and other linguistic objects.

> This harsh ontological banishment from the world of ideals does not apply to compounds built from word types and other linguistic types such as prefixes, and so on. These I take to be abstract structures containing words as constituents. My creationism about words does not extend to sentences. The world in which sentences and other compounds live is brimming with untokened types. Put roughly, the basic elements of the language are earthly creations, but the compounds generated by syntactical rules (the rules also being earthly creations

and thus subject to change) are structures-types-which may or may not have tokens. (511)

The later Kaplan and I are in agreement on the structural nature of sentences and syntactic rules but I do not see why these ought to occupy a different plane of existence to words. In fact there are linguistic reasons for rejecting such a move. We've seen these reasons in the previous section.

11. Kaplan's (1990) account above is a notable exception.

12. Presumably there are two notions of representation at work here. Tokens represent types in a different manner to which types represent their referents. For instance, tokens might 'stand proxy' for types but it doesn't seem to be the case that types are proxies for their referents. The type *horse* does not stand proxy for horses.

13. It should be said that Szabó thinks we disagree on the nature of representation (personal correspondence). He specifically disagrees with me that the architect's blueprints represent something like a building structure-type since if the blueprint were lost and another architect recreated the exact building by pure coincidence, for him, the blueprint would not count as representing that building.

14. There are differences between the two objects. For instance, the equator is a mathematical feature of an object shaped in a certain way. One could argue that its existence is necessary or at least necessarily dependent on the existence of that entire object. National borders, on the other hand, are political objects defined by treaties and historical contingencies. Nevertheless, once they are so conceived they can take on mathematical properties.

15. Hawthorne and Lepore's objection to Kaplan also militates against this suggested distinction in terms of the generative capacity of derivational morphology. To take this idea a bit further, the line between word and sentence seems to be quite clear in so-called *isolating* languages such as English or Chinese but this is not the case for certain members of *agglutinating* and *polysynthetic* language families, where what we would render with an entire sentence can be expressed with a single unit (with affixes appropriately appended). For example, in some of the Yupik languages of parts of Alaska and Siberia larger expression types typically involve a one word root upon which various suffixes are added in order to create sentences or sentence-like structures. Consider an example from Steven Jacobson (1984) of the Central Yupik word *angyaliurvigpaliciquq*, which can be translated as 'he will build a big place for working on boats'. Even Haspelmath (2011), who is generally pessimistic about the possibility of a robust definition of wordhood, thinks the best candidate is the morphosyntactic or structural unit (as opposed to phonetic, semantic or orthographic options).

16. With the exception of Mallory (2020) who places his view squarely within the naturalistic camp. He, like me, advocates that words are not really objects in the ordinary sense. However, he opts for an action-theoretic approach in which tokens provide instructions for the performance of action-types where our normal understanding of 'word' is to be identified with those types. He considers our views to be compatible. I agree in principle, although I do find the type-token distinction still unnervingly lurking in his text.

Chapter 7

1. In terms of (1), given significant theory change, the fruitfulness of the enterprise and its erstwhile discoveries are inevitably called into question. See Stokhof and van Lambalgen 2011, Lappin et al. 2000, Jackendoff 2002.

2. For anyone who might doubt the alacrity of theory change in generative linguistics, I direct them to Chomsky's seminal 1995 book introducing the Minimalist Program. Each chapter therein jettisons constructs carefully argued for in its predecessor. Chomsky (1995: 9) notes that "[w]hat looks reasonable today is likely to take a different form tomorrow. That process is reflected in the material that follows". He goes on to say that "[c]oncepts and principles regarded as fundamental in one chapter are challenged and eliminated in those that follow." These include D-structure, S-structure, government, the θ-criterion, X-bar generally, and the all encompassing Move α operation.

3. There are such instrumentalist theories on the market. See van Fraasen (1980) constructive empiricism as one prominent example. A general problem for such views is that they tend to make miraculous the explanatory and predictive successes of scientific theory. Van Fraasen's response to these sorts of worries is to appeal to an analogy with evolutionary theory such that only the fittest theories survive (where 'fittest' means something like 'latching on to actual regularities in nature') (van Fraasen, 1980: 40). Thanks to Steven French for drawing my attention to this.

4. In Section 7.3, we discuss Ladyman's (2011) account of the structural continuity of the otiose phlogiston theory more closely.

5. For a thorough discussion of the influence of Post on generative grammar, see Pullum (2011) and Lobina (2017).

6. I attempt to follow Pullum and Scholz (2007) throughout in slaloming my way through the minefield of the distinctions between 'formalisation', 'formal', and 'Formalism'. The senses expressed here are related to 'formal' as a term used for systems which abstract over meaning and 'formalisation' as a tool for converting statements of theory into precise mathematical representations. Early generative grammar can be seen as a theory which aimed to achieve both distinct goals.

7. Of course, the term dates back to Lenneberg (1967) who introduced these issues to the generative linguistics community.

8. Matters are not as simple as suggested here. As Bickerton (2014a) stresses, the peculiarity of the situation in linguistics is that the field at present still contains scholars working in various versions of the generative programme concurrently.

9. Basically, regular grammars can't handle constructions like centre embeddings such as *The boy the girl loved left*. These latter constructions form part of a larger class of non-serial dependencies which are inaccessible to regular languages.

10. 'Ordering paradoxes' here refer to the situation in which there are equally valid reasons for orderings from X to Y and Y to X despite the grammar requiring a particular order to pertain.

11. I more or less follow the standard story here but see Kornai and Pullum (1990) for a series of convincing arguments to the effect that the X-bar formalism lacks substance in terms of illuminating phrase structure properties without significant restructuring (which they provide).

NOTES 213

12. There are some linguists who resist this claim. Instead they claim theoretical continuity between the programmes. For instance, Hornstein (2009) offers two reasons for the theoretical continuity between Minimalism and GB.

> First, MP starts from the assumption that GB is roughly correct. It accepts both the general problems identified for solution (e.g. Plato's Problem) and the generalizations ("laws") that have been uncovered (at least to a good first approximation). The second way that MP continues the GB program is in its identification with the Rationalist research strategy that sits at the core of Chomskyan enterprise in general and GB in particular (178).

This might indeed be the case but in my view can best be described as a theoretical orientation rather than theoretical commitment. Many very different theories can be described as 'rationalist' in this broad sense. I also worry about the veracity of the first reason but further discussion will take us into exegetical territory.

13. Technically, as Langendoen (2003) notes "*Merge* is not a single operation, but a family of operations. To belong to the *merge* family, an operation must be able to yield an infinite set of objects from a finite basis" (307). However, by this definition, the phrase structure rules with recursive components would also be included. The structural similarities of various versions of this infinity requirement on grammars will be discussed in the next section.

14. The practice of taking ideas or insights in some disguised form from early frameworks is not uncommon. For example, the binding theory of Government and Binding is very close (if not identical) to principles governing anaphora (like the Ross-Langacker constraints) that were first articulated in the 1960s. Similarly, the trace theory of movement is closely tied to the earlier idea of global derivational constraints. I thank Michael Kac for this observation.

15. Compare this metaphorical language to a similar caution in Pullum (2013: 496), "[t]he fact that derivational steps come in a sequence has encouraged the practice of talking about them in procedural terms. Although this is merely a metaphor, it has come to have a firm grip on linguists thinking about syntax".

16. Of course, the immediate predecessor of phases can be found in *barriers*. See Chomsky (1986) for more details on the general framework.

17. As always things are much more complicated than this. It is often said that the epistemic view is committed to 'hidden' objects or allows a certain agnosticism about the reality of objects. But as French (2014) has argued, for epistemic structural realism to incorporate quantum mechanics it must accept certain metaphysical consequences of certain symmetries (e.g., rotational symmetries) which forces one's hand on the rejection of objects, whether they are hidden or otherwise.

18. At this point, one can glean how such a picture might enter into the debate concerning the ontological foundations of linguistics mentioned earlier. Unlike Platonists who claim among other things that languages are individual abstract objects like sets or mentalists who claim they are psychological or internal states of the brain, a structuralist might argue that languages are complex structures in part identified by abstract rules and physical properties. See Nefdt (2018a) for a similar view.

19. Semi-decidability would work for recursive enumerability as well. For instance, first order logic is not decidable but its validity is recursively enumerable (although I should add that the complement of the validity problem, i.e., determining whether a given formula ϕ is not valid, is not recursively enumerable).

20. He goes on to 'suspect' that the adoption of the derivational approach is more than expository and might indeed be 'correct'.

21. This scenario is guaranteed by Beth's theorem which states (of classical logic) that a non-logical term T is *implicitly* defined by the theory (or generated by the rules) iff an *explicit* definition of the term is deducible from the theory (as in the case of constraint-based or model-theoretic grammars). This effectively connects the proof theory of the logic to the model theory.

22. The literature of WH-movement, for instance, is vast and can be found is almost all textbooks on syntax. Interestingly, for our purposes, the early trace theory is structurally identical to the later Minimalist copy theory of movement. The latter serves an additional theoretical purpose of limiting the proliferation of objects in the ontology such as the indices required for traces.

23. Molecular structures in chemistry have been considerably studied in terms of symmetry and therefore by means of group theory. See Bunker and Jensen (1998) for a comprehensive overview.

24. Cayley (1878, 1889) showed that every group of order n can be represented by a strongly connected digraph of n vertices. There are a number of formal proposals aimed at doing better than Cayley graphs for the representation of group properties. Bretto et al. (2007) define G-graphs which they argue improves upon some of the limitations of Cayley graphs (like the limited information of groups it provides). Vasantha Kandasamy and Smarandache (2009) defines the notion of an identity graph to define identity in groups and expand on some of their other properties. I'll leave it to the eager reader to evaluate those options.

25. Graphs are *locally finite* if each vertex is contained in a finite number of edges, only if the valence of every vertex is finite.

26. Leitgeb and Ladyman (2008) use graph theory to argue for mathematical structures that violate the weak reading of Principle of the Identity of Indiscernibles or "the claim that there are no two distinct individuals that do not stand in some irreflexive relation to each other (other than the relation of non-identity)" (388). They claim that philosophical reflection on mathematical practice in graph theory reveals that the identity (or difference) of places in structures do not outstrip the structures in which they are found. Leitgeb (2020) goes a step further to mount a defence of non-eliminative structuralism based on graph theory, specifically unlabelled graphs. If you assign numbers (or some other label) to the nodes of a graph, it is labelled. Unlabelled graphs have no such identifiers, except in terms of their connections to other nodes. He develops an axiomatic theory of unlabelled graphs to show that graphs are *sui generis* structures not necessarily represented in terms of set theory. After a survey of a few characterisations of unlabelled graphs, Leitgeb states:

> Clearly, these mathematical texts present unlabeled graphs informally in much the same way in which structuralists talk about mathematical structures [...] Benacerraf's 'To *be* the number 3 is no more or less than to be preceded by 2, 1,

and possibly 0, to be followed by 4, 5, and so forth', [...] resembles the 'individual nodes have no distinct identifications except through their interconnectivity'. (2020: 329)

27. The universal invariant of labelled graphs is given by unlabelled graphs. From a structural perspective, if the number of nodes matches the labels then we can say that two labelled graphs are isomorphic if the corresponding underlying unlabelled graphs are isomorphic. Thus, the noneliminative structuralism advocated for unlabelled graphs by Leitgeb can be extended to at least a subset of the kinds of labelled graphs linguists typically use, at least in principle.

28. See Michaelis (2001); Mönnich (2007) or rather Stabler's (1997) formalisation and interpretation of minimalist syntax has been shown to be equivalent to some of the above grammars.

29. In a sense, 'formal language' is polysemous in referring to a meaningless or uninterpreted calculus akin to its use in the Hilbert programme and a component of a compositional logical language assumed in the work of Frege and Russell. See van Heijnoort (1967) on the history of the different views.

30. One possible candidate for how to identify this structure can be found in Stabler (2019). There he motivates the formal structure of tree languages as point of connection between substitution classes (transformations), dependencies and constraint-based formalisms.

31. See Buldt et al. (2008) for an alternative epistemological foundation for mathematics along similar lines.

Chapter 8

1. It's also popped up in various guises throughout philosophy, in Carnap (1950) and Quine (1953), in computer science in Dijkstra (1968), and in cognitive science in Marr (1982).

2. Chomsky's work in phonology and morphology is equally foundational to the science. I thank Gabe Dupre for reminding me of this fact.

3. Their third truism or the claim that 'complexity can come from simplicity' is also something dear to the hearts of many linguists across theoretical divides. Minimalists with their claims concerning the centrality of the *merge* operation are one case. On the computational experimental side, 'iterated learning models' in language evolution research aim to explain how complex syntactic structure, such as discrete infinity, can be generated by creating highly simplistic models involving generational simulations of populations with no language to begin with (Brighton and Kirby, 2001, see). Even in the semantic evolution literature, compelling accounts concerning how the emergence of complex meaning might have started out as simple Non-Gricean expressive communication similar to animal signals have been proffered, see Bar-On (2013, 2017).

4. Despite the name, OOP is in clear violation of the type-token distinction as the object model and the real world systems it tracks are never isomorphic, thus the representation is indirect.

5. Floridi and Sanders (2004) also insist on the discreteness of observables which renders their possible values finite as opposed to the case with analogue systems.

6. There are some further technicalities present in both Floridi and Sanders (2004) and Floridi (2008b) that I've elided here for the sake of accessibility. I think the picture is clear enough but an interested reader is referred to the aforementioned works for more formal detail.

7. There is a wealth of literature on this principle. I can't possibly do it justice here (see Nefdt, 2020b for a metasemantic approach and 2020a for application to issues in AI and NLP). But the basic idea is that some form of the following dictum is true: [t]he meaning of a complex expression is a function of the meanings of its constituents and the syntactic rule used to combine these constituents.

8. As Müller notes, later versions of Minimalism allow Spell-Out at multiple points in the derivation. But we won't need that detail here.

9. Chomsky (2015: 152) himself explicitly endorses this picture:

> A standard assumption is that [Universal Grammar] specifies certain *linguistic levels*, each a symbolic system, often called a "representational system". Each linguistic level provides the means for presenting certain systematic information about linguistic expressions. Each linguistic expression [structural description] is a sequence of representations, one at each linguistic level.

10. This situation is compounded by the fact that the PA recognises further interface rules within individual systems or 'tiers' (cf. LFG below).

11. As Börjars (2020) notes, "in comparison to other frameworks, LFG's approach to X-bar syntax is unorthodox in that, for instance, nonbinary branching as well as exocentric categories is permitted" (157) and "that all nodes, including preterminal and head nodes, are optional" (159).

12. Teleosemantic accounts like Millikan (1984) would be interpretationist by this standard and Shea's (2020) recent account would be more in line with productivism.

13. See Pietroski (2018) for a recent formulation of internalist semantics.

14. Chomsky has also claimed that there are certain cases of co-predication in which these meanings can coalesce. See Ortega-Andrés and Vicente (2018) for a recent overview of the phenomenon.

15. See Dummett 1981, Hale 1997, Pelletier 2001 for some useful discussion.

16. This opens the door to a possible formal metasemantics in which the mereological principles governing metastructures are determined similar to Shapiro's (1997) formal framework for structures based on the axioms of set-theory.

17. Although related, model-theoretic syntax takes a sentence to be a model of a grammar iff it meets certain conditions on grammaticality. This is different from the application of model theory in semantics.

18. This is only in the case of a simplified extensional account which is sufficient for the present point.

19. For a top-down approach to compositionality, see Brandom (2007) on inferentialism. And for an argument that inferentialism is a kind of structuralism, see Peregrin (2008) and Nefdt (2018b).

20. I haven't said anything here about proof-theoretic semantics but I think given its shared mathematical foundations with syntax, it naturally aligns with a structuralist picture.

21. Stanton (2020), in the same volume, goes one step further to assert that these approaches might cause significant disruption to the *status quo* in semantics and as a result, its 'special relationship' with philosophy.

22. Dupre (2020) has a related metasemantic argument showing that truth-conditional semantics is scientifically compatible with radical contextualism, i.e., you could successfully model semantics in terms of truth-conditions whether or not they pertain at the metasemantic level.

23. Movement away from discrete structures to distributional patterns or continuous methods might involve infinite values for variables in the LoAs but this can be accommodated in the method of levels of abstraction of Floridi and company with some modifications or one could bring in continuous methods like differential calculus.

Chapter 9

1. By 'formal' here I refer to the mathematical underpinnings of linguistics inspired by Post's developments of canonical production systems in proof-theory. For more information, see Pullum (2011).

2. I avoid the terms 'first' and 'second' here as Chomsky (1991) and Boeckx (2005) claim the first cognitive revolution occurred in the eighteenth century in the form of the Cartesian representational theory of perception.

3. Quine even constrained solutions to his infamous indeterminacy arguments to involving only observable behavioural evidence.

4. Of course, there were many more components of this result including Bar-Hillel's work on applying recursion theory to syntax, Harris' transformations and Goodman's constructional systems theory. See Tomalin (2006) for a detailed review. See also Pullum and Scholz (2007) for requisite detail on Post's influence neglected by Tomalin's account.

5. Although it should be noted that clinical psychology of the time largely resisted the behaviouristic scruples of experimental psychology.

6. This date is corroborated by Gardner (1985).

7. A more full account would include reference to the work of Vygotsky and Luria in the Soviet Union among other things.

8. The formal grammars of generative grammar correspond to various automata in accordance with the Chomsky Hierarchy. For instance, context free grammars can be represented by pushdown automata while regular grammars only map onto a more restrictive class of finite state automata. See Chapter 4 for some details.

9. See Proudfoot and Copeland (2012) for more on this shift.

10. This view is the subject of the philosophical critique of Devitt (2006) in the philosophy of linguistics and also the more recent so-called 4E approach to cognitive science represented by embodied, embedded, extended and enacted cognition. In fact, Menary (2010) argues that perhaps one of the only things connecting these latter views is their mutual rejection of cognitivism or representationalism.

11. Some people might think that the second generation was merely a harbinger of the fate of the enterprise as a whole. In a controversial paper, Núñez et al. (2019) make the case for the failure of both the interdisciplinary and integrational elements of cognitive science, thus sparking the 'cognitive science is dead' debate (see the special issue of *Topics in Cognitive Science* (2019), Vol. 11 for a robust discussion). I don't plan to get involved in that polemic here. But the present work does indirectly address some of the core worries of the original Núñez et al. article and even if my arguments are directed at a ghost, perhaps they could at least convince the reader that a séance is in order. I thank Edouard Machery for impressing the relevance of this issue upon me.

12. This last field has seen a resurgence of interest in the connection between language and thought exemplified by the erstwhile Sapir-Whorf hypothesis. See Reines and Prinz (2009) for an overview of this literature.

13. For instance, Langendoen and Postal (1984) provided an argument for the cardinality of natural language being that of a proper class.

14. Of course, mentalism in linguistics was more explicitly established after some of these procedural mechanisms were incorporated in the theory.

15. See Cappelen (2017) for a negative view on the relationship between philosophy and formal linguistics, especially semantics and Nefdt (2019d) for a response.

16. See Hintikka (1999) for one such criticism, and Maynes and Gross (2013) for an overview of the literature.

17. There is further work in probabilistic linguistics on the potential connections between formal grammars and statistical Bayesian approaches. See Bod et al. (2003) for an overview of the field.

Bibliography

Ainsworth, P. 2010. What is ontic structural realism? *Studies in History and Philosophy of Science Part B*, 41(1): 50–57.

Anderson, P. 1972. More is different. *Science*, 177(4047): 393–396.

Anderson, S., & Lightfoot, D. 2002. *The Language Organ*. Cambridge University Press.

Ortega-Andrés, M. & Vicente, A. 2018. Polysemy and co-predication. *Glossa: A Journal of General Linguistics* 4(1), p. 1. doi: https://doi.org/10.5334/gjgl.564.

Awodey, S. 1996. Structure in mathematics and logic: A categorical perspective. *Philosophia Mathematica*, (3)4: 209–237.

Bach, E & Miller, P. 2003. Generative capacity. In William J. Frawley (ed.). *International Encyclopedia of Linguistics* (2nd ed.), pp. 20–21. Oxford University Press.

Baltzly, D. 1996. Stoicism. *The Stanford Encyclopedia of Philosophy*, Zalta, E. (ed.), https://plato.stanford.edu/archives/spr2019/entries/stoicism/.

Barnes, J., S. Bobzien, & Mignucci, M. 1999. Logic. In Algra, K., Barnes, J. Mansfeld, J., & Schofield, M. (eds.), *Hellenistic Philosophy*, pp. 77–176. Cambridge University Press.

Bar-On, D. 2013. Origins of meaning: Must we 'go Gricean'? *Mind & Language*, 28(3): 342–375.

Bar-On, D. 2017. Communicative intentions, expressive communication, and origins of meaning. In Andrews, K. & Beck, J. (eds.) *The Routledge Handbook of Philosophy of Animal Minds*, pp. 301–312. Routledge.

Bartlett, F. 1932. *Remembering: A study in experimental and social psychology*. Cambridge University Press.

Batterman, R. 2002. *The Devil in the Details*. Oxford University Press.

Bechtel, W., Abrahamsen, A., & Graham, G. 2001. Cognitive Science: History. In *International Encyclopedia of the Social & Behavioral Sciences*, pp. 2154–2158. Elsevier.

Benacerraf, P. 1965. What numbers could not be. *Philosophical Review*, 74: 47–73.

Benacerraf, P. 1973. Mathematical truth. *The Journal of Philosophy*, 70(19): 661–679.

Beni, M. 2019a. Conjuring cognitive structures: Towards a unified model of cognition. In Nepomuceno-Fernández Á., Magnani L., Salguero-Lamillar F., Barés-Gómez C., Fontaine M. (eds.) *Model-Based Reasoning in Science and Technology*. Springer.

Beni, M. 2019b. *Cognitive Structural Realism: A radical solution to the problem of scientific representation*. Springer.

Bernays, P. 1923. Erwiderung auf die Note von Herrn Aloys Müller: Über Zahlen als Zeichen. *Mathematische Annalen*, 90:159–63, 1923. English translation in [Mancosu, 1998a, 223–226].

Berwick, R., & Chomsky, N. 2016. *Why only us? Language and evolution*. Cambridge, MA: The MIT Press.

Berwick, R., Okanoya, K., Beckers, G., & Bolhuis, J. 2011. Songs to syntax: The linguistics of birdsong. *Trends in Cognitive Sciences*, 15(3): 113–121.

Bickerton, D. 2014a. *More than nature needs: Language, mind, and evolution*. Harvard University Press.

Bickerton, D. 2014b. Problems with biolinguistics. *Biolinguistics*, 8: 73–96.

Blackburn, P. 2006. Arthur Prior and hybrid logic. *Synthese* 150(3): 329–372.

Bloomfield, L. 1936. Language or ideas? *Language* 12: 89–95.

Blutner, R. 2011. Taking a broader view: Abstraction and idealization. *Theoretical Linguistics*, 37: 1–2.

Boas, H., & Sag, I. (eds.). 2012. *Sign-based Construction Grammar*. Stanford.

Bod, R., Hay, J., & Jannedy, S. 2003. *Probabilistic Linguistics*. Cambridge, MA: The MIT Press.

Boeckx, C. 2005. Generative grammar and modern cognitive science. *Journal of Cognitive Science* 6: 45–54.

Boeckx, C. 2006. *Linguistic Minimalism: Origins, concepts, methods, and aims*. Oxford: Oxford University Press.

Boeckx, C., & Martins, P. 2016. Biolinguistics. *Oxford Research Encyclopedias, Linguistics*, pp. 1–13.

Bonolis, L. 2004. From the rise of the group concept to the stormy onset of group theory in the new quantum mechanics: A saga of the invariant characterization of physical objects, events and theories. *Rivista Del Nuovo Cimento* 27: 4–5.

Boogerd, F., Bruggeman, F., Hofmeyr, J., & Westerhoff, H. (eds.). 2007. *Systems Biology: Philosophical foundations*. Amsterdam: Elsevier.

Boolos, G. 2000. Must we believe in Set Theory. In Sher, G., and Tieszen, R. (eds.) *Between Logic and Intuition: Essays in honor of Charles Parsons*. Cambridge University Press.

Börjars, K. 2020. Lexical-Functional Grammar: An Overview. *Annual Review of Linguistics*, 6(1): 155–172.

Boroditsky, L., Schmidt, L., & Phillips, W. 2003. Sex, Syntax and Semantics. *Language in Mind: Advances in the Studies of Language and Cognition*. Ed. Gentner and Goldin-Meadow. Cambridge, MA: The MIT Press.

Brandom, R. 1994. *Making It Explicit*. Harvard University Press.

Brandom, R. 2007. Inferentialism and some of its challenges. *Philosophy and Phenomenological Research*, 74(3): 651–676.

Bretto, A., Faisant, A., & Gillibert, L. 2007. G-graphs: A new representation of groups. *Journal of Symbolic Computation* 42: 549–560.

Bresnan, J., Asudeh, A., Toivonen, I., Wechsler, S. 2016. *Lexical-Functional Syntax*. Wiley.

Brighton, H. & Kirby, S. 2001. The survival of the smallest: Stability conditions for the cultural evolution of compositional language. In Kelemen, J., and Sosik, P., (eds.), *Advances in Artificial Life*, pp. 592–601. Springer.

Bromberger, S. 1989. Types and tokens in linguistics. In George, A., *Reflections on Chomsky*, pp. 58–90. Basil Blackwell.

Bromberger, S. 2011. What are words? *The Journal of Philosophy*, 108(9): 486–503.

Büchi, R. 1960. Weak second-order arithmetic and finite automata. *Zeitschrift für mathematische Logik und Grundlagen der Mathematik* 6: 66–92.

Buckner, C. 2019. Deep learning: A philosophical introduction. *Philosophy Compass*, 14(10): e12625.

Bueno, O., & Colyvan, M. 2011. An inferential conception of the application of mathematics. *Nous* 45(2): 345–374.

Buldt, B., Löwe, B., & Müller, T. 2008. Towards a new epistemology of mathematics. *Erkenntnis*, 6: 309–329.

Bunker, P., & Jensen, P. 1998. *Molecular Symmetry and Spectroscopy*, 2nd Edition. NRC Press.

Burgess, A., & Sherman, B. 2014. Introduction. In Burgess, A., & Sherman, B., (eds.) *Metasemantics: New essays on the foundations of meaning*. Oxford University Press.

Burten-Roberts, N., & Poole, G. 2006. Virtual conceptual necessity, feature-dissociation and the Saussurian legacy in generative grammar. *Journal of Linguistics*, 42: 575–628.

Cappelen, H. 2017. Why philosophers should not do semantics. *Review of Philosophy and Psychology*, 8(4): 743–762.

Cappelen, H. 2018. *Fixing Language: An essay on conceptual engineering*. Oxford: Oxford University Press.

Carnap, R. 1950. Empiricism, semantics, and ontology. *Revue Internationale de Philosophie*, 4: 20–40.

Carruthers, P. 2006. *The Architecture of Mind: Massive modularity and the flexibility of thought*. Oxford University Press.

Cayley, A. 1878. The theory of groups: Graphical representations. *American Journal of Mathematics* 1: 174–176.

Cayley, A. 1889. On the theory of groups. *American Journal of Mathematics* 11: 139–157.

Chalmers, D. 2011. A computational foundation for the study of cognition. *The Journal of Cognitive Science*, 12: 323–357.

Chihara, C. 2004. *A structural account of mathematics*. Oxford University Press.

Chipman, S. 2017. An introduction to cognitive science. In Chipman, S. (ed.) *The Oxford Handbook of Cognitive Science*, pp. 1–22. Oxford University Press.

Chomsky, N. 1956. Three models for the description of language. *IRE Transactions on Information Theory*, 2: 113–123.

Chomsky, N. 1957. *Syntactic structures*. The Hague: Mouton.

Chomsky, N. 1959. Review of Skinner's *Verbal Behavior*. *Language*, 35: 26–58.

Chomsky, N. 1963. Formal properties of grammars. In Duncan, L., Bush, R., Galanter, E (eds.). *Handbook of Mathematical Psychology*, pp. 323–418. II. John Wiley and Sons.

Chomsky, N. 1965. *Aspects of a Theory of Syntax*. Cambridge, MA: The MIT Press.

Chomsky, N. 1970. Remarks on nominalization. In Jacobs, R., & Rosenbaum, P. (eds.), *Readings in English transformational grammar*, pp. 184–221. Ginn.

Chomsky, N. 1972. *Studies on Semantics in Generative Grammar*. Mouton.

Chomsky, N. 1975. *The Logical Structure of Linguistic Theory*. Springer.

Chomsky, N. 1980. *Rules and Representations*. Columbia University Press.

Chomsky, N. 1986. *Knowledge of Language: Its nature, origin, and use*. Praeger.

Chomsky, N. 1988. *Language and Problems of Knowledge: The Managua lectures*. Cambridge, MA: The MIT Press.

Chomsky, N. 1991. Linguistics and adjacent fields: A personal view. In Kasher, A. (ed.), *The Chomskyan Turn*, pp. 3–25. Blackwell.

Chomsky, N. 1993. A minimalist program for linguistic theory. In Hale, K., & Keyser, S. (eds.), *The View from Building 20: Essays in Honor of Sylvain Bromberger*, pp. 1–53. Cambridge, MA: The MIT Press.

Chomsky, N. 1995. *The Minimalist Program*. Cambridge, MA: The MIT Press.

Chomsky, N. 1998. Minimalist inquiries: The framework. *MIT Occasional Papers in Linguistics*, 15. MIT Department of Linguistics.

Chomsky, N. 1999. Derivation by phase. *MIT Occasional Papers in Linguistics*, 18. MIT Department of Linguistics.

Chomsky, N. 2000a. *New Horizons for the Study of Mind and Language*. Cambridge University Press.

Chomsky, N. 2000b. Minimalist inquiries. In Martin, R., Michaels, D., & Uriagereka, J., (eds.), *Step by Step: Essays on minimalist syntax in honor of Howard Lasnik*, pp. 89–155. Cambridge, MA: The MIT Press.

Chomsky, N. 2003. Reply to Millikan. In Anthony, L., & Hornstein, N. *Chomsky and His Critics*, pp. 308–315. Oxford: Blackwell.

Chomsky, N. 2005. Threes factors in language design. *Linguistic Inquiry* 36(1): 1–22.

Chomsky, N. 2008. On phases. In Freiden, R., Otero, C., & Zubizarreta, M. (eds.), *Foundational Issues in Linguistic Theory*, pp. 133–166. Cambridge, MA: The MIT Press.

Christiansen, M., & Chater, N. 2016. The Now-or-Never bottleneck: A fundamental constraint on language. *Behavioural and Brain Sciences*, 39: e62.

Clark, H. 1996. *Using Language*. Cambridge University Press.

Clark, A., & Chalmers, D. 1998. The extended mind. *Analysis*, 58(1): 7–19.

Croft, W. 2013. Radical construction grammar. In Hoffmann, T., & Trousdale, G., *The Oxford Handbook of Construction Grammar*. Oxford University Press.

Collins, J. 2007. Review of Devitt 2006a. *Mind*, 116: 416–423.

Collins, J. 2008a. Knowledge of language redux. *Croatian Journal of Philosophy*, 8: 3–43.

Colombo, M., & Wright, C. 2021. First principles in the life sciences: The free-energy principle, organicism, and mechanism. *Synthese*, 198: 3463–3488.

Corballis, M. 2007. Recursion, language, and starlings. *Cognitive Science*, 31: 697–704.

Cowie, F. 1999. *What's Within? Nativism Reconsidered*. Oxford University Press.

Culicover, P. 2011. Core and periphery. In Hogan, P., (ed.), *The Cambridge Encyclopedia of the Language Sciences*, pp. 227–230. Cambridge University Press.

Dalrymple, M., & Kehler, A. 1995. On the constraints imposed by 'respectively'. *Linguistic Inquiry*, 26: 531–536.

Davidson, D. 1965. Theories of meaning and learnable languages. Reprinted in *Inquiries into Truth and Interpretation*. Clarendon Press, 2001, 3–16.

Davidson, D. 1986. A nice derangement of epitaphs. In Davidson, D., (ed.) *Truth and Interpretation*, pp. 251–265. Blackwell.

Davidson, D. 2001. The emergence of thought. In Davidson, D., (ed.) *Subjective, Intersubjective, Objective*, pp. 123–134. New York: Oxford University Press.

Debusmann, R. 2006. *Extensible Dependency Grammar: A Modular Grammar Formalism Based On Multigraph Description*. PhD Thesis.

Dennett, D. 1971. Intentional systems. *The Journal of Philosophy*, 68(4): 87–106.

Dennett, D. 1987. True believers. In Dennett, D., (ed.) *The Intentional Stance*. Cambridge, MA: The MIT Press.

Dennett, D. 1991. Real patterns. *Journal of Philosophy*, 88: 27–51.

Dennett, D. 1996. *The Intentional Stance*. Cambridge, MA: The MIT Press.

De Vignemont, F. 2018. The extended body hypothesis: Referred sensations from tools to peripersonal space. In A. Newen, L. De Bruin, & S. Gallagher (eds.), *The Oxford handbook of 4E cognition*. Oxford University Press.

Devitt, M. 2006. *Ignorance of Language*. Oxford University Press.

Devitt, M. 2008. Explanation and reality in linguistics. *Croatian Journal of Philosophy*, 8(23): 203–223.

Devitt, M. 2013. The "linguistic conecption" of grammars. *Filozofia Nauki*, 2(82): 5–14.

Dhillon, B., Smith, M., Baghela, A., Lee, A., & Hancock, R. 2020. Systems biology approaches to understanding the human immune system. *Frontiers of Immunology*, 11: 1683.

Diessel, H. 2013. Construction grammar and first language acquisition. In Hoffmanm, T., & Trousdale, G. (eds.) *The Oxford Handbook of Construction Grammar*, pp. 347–364. Oxford University Press.

Dijkstra, E. 1968. The structure of THE multiprogramming system. *Communications of the ACM*, 11(5): 341–346.

Dipert, R. 1997. The mathematical structure of the world: The world as graph. *The Journal of Philosophy*, 94(7): 329–358.

Dummett, M. 1981a. *Frege: Philosophy of Language*. Harvard University Press.

Dummett, M. 1993. What do I know when I know a language? In Dummett, M. (ed.) *The Seas of Language*. Clarendon Press.

Dupre, G. 2020. Idealisation in semantics: Truth-conditional semantics for radical contextualists. *Inquiry* (online first).

Dupre, G. (forthcoming). Classifying data and interpreting theories: The case of generative grammar. In Bueno, O. & Martinez-Ordaz, M. (eds.), *From Contradiction to Defectiveness to Pluralism in Science: Philosophical and Formal Analyses* (Synthese Library).

Dupré, J., & O'Malley, M. 2007. Metagenomics and biological ontology. *Studies in History and Philosophy of Biological and Biomedical Sciences*, 38(4): 834–846.

Dupré, J., & O'Malley, M. 2009. Varieties of living things: Life at the intersection of lineage metabolism. *Philosophy and Theory in Biology*, 1(3): 1–25.

Erk, K. 2020. Variations on abstract semantic spaces. In Nefdt, R., Klippi, C., & Karstens, B., (eds.) *The Philosophy and Science of Language*, pp. 79–99. Palgrave Mcmillan.

Epstein, W., & Hatfield, G. 1994. Gestalt psychology and the philosophy of mind. *Philosophical Psychology* 7(2), 163–181.

Fillmore, C., Paul, P., & O'Connor, M. 1988. Regularity and idiomaticity in grammatical constructions: The case of *let alone*. *Language*, 64: 501–538.

Firth, J. 1957. *Papers in Linguistics*. Oxford University Press.

Floridi, L. 2008a. A defence of informational structural realism. *Synthese*, 161(2): 219–253.

Floridi, L. 2008b. The method of levels of abstraction. *Minds & Machines*, 18: 303–329.

Floridi, L. 2011. *The Philosophy of Information*. Oxford University Press.

Floridi, L., & Sanders, J. 2004. The method of abstraction. In Negrotti, M. (ed.) *Yearbook of the Artificial: Nature, culture, and technology. Models in contemporary sciences*, pp. 177–220. Peter Lang.

Fodor, J. 1974. Special sciences, or the disunity of science as a working hypothesis. *Synthese* 28: 77–115.

Fodor, J. 1975. *The Language of Thought*. Thomas Y. Crowell.

Fodor, J. 1981. *Representations*. Cambridge, MA: The MIT Press.

Fodor, J. 1987. *Psychosemantics*. Cambridge, MA: The MIT Press.

Fodor, J., Bever, T., & Garrett, M. 1974. *The Psychology of Language*. McGraw-Hill Publishers.

Fodor, J., & Piattelli-Palmarini, M. 2010. *What Darwin Got Wrong*. Farrar, Straus & Giroux.

Frankish, K., & Ramsey, W (eds.). 2012. *The Cambridge Handbook of Cognitive Science*. Cambridge University Press.

Frege, G. 1879. *Begriffsschrift*. Louis Nebert; trans. by S. Bauer-Mengelberg, in J. van Heijenoort, ed., *From Frege to Gödel*. Harvard, 1967.

Frege, G. 1893. *Grundgesetze der Arithmetik*, vol. 1, 1893, Jena; trans. by M. Furth as *The Basic Laws of Arithmetic*. University of California Press, 1964.

Frege, G. 1953. *Foundations of Arithmetic*, trans. by J. L. Austin. Blackwell.

Freidin, R. 2012. A brief history of generative grammar. In Russell, G., & Fara, D., (eds.), *The Routledge Companion to Philosophy of Language*, pp. 895–916. Routledge.

French, S. 1999. Models and mathematics in physics: The role of group theory. In Butterfield, J., & Pagonis, C., (eds.), *From Physics to Philosophy*, pp. 187–207. Cambridge: Cambridge University Press.

French, S. 2000. The reasonable effectiveness of mathematics: Partial structures and the application of group theory to physics. *Synthese* 125 (1–2): 103–20.

French, S. 2006. Structure as a weapon of the realist. *Proceedings of the Aristotelian Society*, 106: 167–185.

French, S. 2011. Shifting the structures in physics and biology: a prophylactic promiscuous realism. *Studies in History and Philosophy of Biological and Biomedical Sciences*, 42: 164–173.

French, S. 2014. *The Structure of the World*. Oxford University Press.

French, S., & Krause, D. 2006. *Identity in Physics: A Historical, Philosophical, and Formal Analysis*. Oxford University Press.

French, S., & Ladyman, J. 2003. Remodelling structural realism: quantum physics and the metaphysics of structure. *Synthese*, 136: 31–56.

French, S., & Redhead, M. 1988. Quantum physics and the identity of indiscernibles. *British Journal for the Philosophy of Science*, 39(2): 233–246.

Frigg, R., & Votsis, I. 2011. Everything you always wanted to know about structural realism but were afraid to ask. *European Journal for Philosophy of Science* 1: 227–276.

Gao, J., Li, D., & Havlin, S. 2014. From a single network to a network of networks. *National Science Review*, 1(3), 346–356.

Ganascia, J. 2015. Abstraction of levels of abstraction. *Journal of Experimental & Theoretical Artificial Intelligence*, 27(1): 23–35.

Gardner, H. 1985. *The Mind's New Science: A History of the Cognitive Revolution*. Basic Books.

Gärdenfors, P. 2000. *Conceptual Spaces: The geometry of thought*. Cambridge, MA: The MIT Press.

Gasparri, L. 2020. A pluralistic theory of wordhood. *Mind & Language*, 36(4): 592–609.

Gasser, M. 2015. Structuralism and its ontology. *Ergo*, 2(1): 1–26.

Gazdar, G., Klein, E., Pullum, G., & Sag, I. 1985. *Generalized Phrase Structure Grammar*. Harvard University Press.

Gentner, T., Fenn, K., Margoliash, D & Nusbaum, H. 2006. Recursive syntactic pattern learning by songbirds. *Nature*, 440(7088): 1204–1207.

Giere, R. 1988. *Explaining Science: A cognitive approach*. Chicago University Press.

Godfrey-Smith, P. 2006. The strategy of model-based science. *Biology and Philosophy*, 21: 725–740.

Goldberg, Adele. 2013. Constructionist approaches. In Hoffmann, T., & Trousdale, G., *The Oxford Handbook of Construction Grammar*. Oxford University Press.

Goldberg. A. 2015. Compositionality. In Reimer, N. (ed.) *Routledge Semantics Handbook*, pp. 419–433. Routledge Press.

Greco, G., Paronitti, G., Turilli, M., & Floridi, L. 2005. How to do philosophy informationally. *Lecture Notes in Computer Science*, 3782: 623–634.

Greenberg, J. 1963. Some universals of grammar with particular reference to the order of meaningful elements. In Greenberg, J. (ed.), *Universals of Language*, pp. 73–113. Cambridge, MA: The MIT Press.

Grice, P. 1957. Meaning. *The Philosophical Review* 66: 377–388.

Hacker, P. 2014. Two conceptions of language. *Erkenntnis*, 79(7): 1271–1288.

Hacking, I. 1983. *Representing and intervening*. Cambridge University Press.

Hale, B. 1997. Grundlagen 64. *Proceedings of the Aristotelian Society* 97: 243–261.

Harat, M., Rudas, M., & Rybakowski, J. 2008. Psychosurgery: the past and present of ablation procedures. *Neuroendocrinology Letters*. Suppl 1: 105–22.

Harris, R. 2021. *Linguistics Wars*. Oxford University Press.

Haspelmath, M. 2011. The Indeterminacy of Word Segmentation and the Nature of Morphology and Syntax. *Folia Linguistica* 45 (1): 31–80.

Hasselman F., Seevinck M., & Cox R. 2015. Caught in the undertow: there is structure beneath the ontic stream. *SSRN Electron Journal*. http://www.ssrn.com/abstract=2553223.

Hauser, M., Hauser, M., & Fitch, W. 2002. The Faculty of Language: What is it, who has it, and how did it evolve? *Science*, 298: 1569–1579.

Hawthorne, J., & E. Lepore. 2011. On Words. *The Journal of Philosophy* 108 (9): 447–485.

Hays, D. 1964. Dependency theory: A formalism and some observations. *Language* 40: 511–525.

Hellman, G. 1989. *Mathematics without Numbers: Towards a Modal-Structural Interpretation*. Oxford: Clarendon Press.

Hellman, G. 2001. Three varieties of mathematical structuralism. *Philosophia Mathematica*, 9(2): 184–211.

Hendry, R. 2021. Structure, scale and emergence. *Studies in History & Philosophy of Sicence Part A*, 85: 44–53.

Herbelot, A., & Copestake, A. 2013. Lexicalised compositionality. https://www.cl.cam.ac.uk/~aac10/papers/lc3-0web.pdf

Hilbert, D. 1899 [2004]. David Hilbert's Lectures on the Foundations of Geometry, 1891–1902. In Majer, U & Hallbett, M (eds.). Springer, New York.

Hintikka, J. 1988. On the development of the model-theoretic viewpoint in logical theory. *Synthese* 77(1): 1–36.

Hintikka, K. 1999. The emperor's new intuitions. *The Journal of Philosophy*, 96(3): 127–147.

Hinzen, W. 2012. Minimalism. In R. Kempson, T. Fernando, & N. Asher (eds.), *Philosophy of linguistics*, pp. 91–141. Oxford: Elsevier B.V North Holland.

Hjelmslev, L. 1953[1943]. *Prolegomena to a Theory of Language*. Baltimore: Indiana University Publications in Anthropology and Linguistics (IJAL Memoir, 7) (2nd OD (slightly rev.): Madison: University of Wisconsin Press, 1961.

Hockett, C. 1958. *A course in modern linguistics*. New York: Macmillan.

Hoffmann, T. & Trousdale, G. 2013. Construction Grammar: Introduction. In Hoffmann, T., & Trousdale, G. (eds.) *The Oxford Handbook of Construction Grammar*. Oxford University Press.

Horstein, N. 2009. *A Theory of Syntax: Minimal Operations and Universal Grammar*. Cambridge University Press.

Hopcroft, J., & Ullman, J. 1979. *Introduction to Automata Theory, Langauge, and Computation*. Addison-Wesley, Reading MA.

Hopcroft, J., Motwani, R., & Ullman, J. 2001. *Introduction to Automata Theory, Langauge, and Computation, 2nd Edition*. Addison-Wesley, Reading MA.

Hopper, P. 1987. Emergent Grammar. *Berkeley Linguistics Society*, 13: 139–157.

Huddleston, R., & Pullum, G. 2002. *The Cambridge grammar of the English language*. Cambridge: Cambridge University Press.

Hudson, Richard. 1990. *English Word Grammar*. Oxford: Basil Blackwell.

Hunter, T. 2021. The Chomsky Hierarchy. In Allott, N. Lohndal, T., & Rey, G. (ed.), *Blackwell Companion to Chomsky*, pp. 74–95. Wiley-Blackwell.

Hutchins, E. (1995). *Cognition in the wild*. Cambridge, MA: The MIT Press.

Hutto, D., & Satne, G. 2017. Continuity scepticism in doubt: A radically enactive take. In C. Christoph Durt, T. Fuchs, & C. Tewes (eds.), *Embodiment, enaction, and culture: Investigating the constitution of the shared world*, pp. 107–127. Cambridge, MA: The MIT Press.

Irmak, N. 2019. An ontology of words. *Erkenntnis* 84(5): 1139–1158.

Itkonen, E. 2013. Philosophy of linguistics. In Allan, K. (ed.), *The Oxford Handbook of the History of Linguistics*, pp. 747–774. Oxford University Press.

Jackendoff, R. 1977. *X' Syntax*. Cambridge, MA: The MIT Press.

Jackendoff, R. 2002. *Foundations of Language: Brain, meaning, grammar, evolution*. Oxford University Press.

Jackendoff, R. 2007. Linguistics in cognitive science: The state of the art. *The Linguistic Review*, 24: 347–401.

Jackendoff, R. 2013. Constructions in the parallel architecture. In Hoffmann, T., & Trousdale, G., *The Oxford Handbook of Construction Grammar*. Oxford University Press.

Jackendoff, R. 2017. In defense of theory. *Cognitive Science*, 41: 185–212.

Jackendoff, R. 2018. Representations and rules of language. In Huebner, B., (ed.) *The Philosophy of Daniel Dennett*, pp. 95–126. Oxford University Press.

Jackendoff, R., & Wittenberg, E. 2014. What you can say without syntax: A hierarchy of grammatical complexity. In Newmeyer, F., & Preston, L. (eds.), *Measuring Linguistic Complexity*, pp. 65–82. Oxford University Press.

Jacobson, S. 1984. *Central Yupi'k and the Schools: A Handbook for Teachers*. Alaska Native Language Center and Alaska Department of Education, alaskool.org.

Jacobson, P. 2014. *Compositional Semantics: An Introduction to the Syntax/Semantics Interface*. Oxford University Press.

Jäger, G., & Rogers, J. 2012. Formal language theory: Refining the Chomsky Hierarchy. *Philosophical Transactions of the Royal Society B: Biological Sciences* 367: 1956–1970.

Jerne, N. 1985. The generative grammar of the immune system. *Science*, 229: 1057–1059.

Johnson, K. 2007. The legacy of methodological dualism. *Mind & Language*, 22(4): 366–401.

Johnson, K. 2015. Notational variants and invariance in linguistics. *Mind & Language*, 30(2): 162–186.

Joseph, J. 1999. How structuralist was "American structuralism"? *Henry Sweet Society Bulletin*, 33: 23–28.

Joseph, J. 2015. Iconicity in Saussure's linguistic work, and why it does not contradict the arbitrariness of the sign. *Historiographia Linguistica*, 42(1): 85–105.

Joseph, J. 2017. Ferdinand de Saussure. *Oxford Research Encyclopedia of Linguistics*, pp. 1–17.

Joshi, A. 1985. How much contextsensitivity is required to provide reasonable structural descriptions: Tree adjoining grammars. In Dowty, D., Karttunen, L., and Zwicky, A., (eds.) *Natural Language Processing: psycholinguistic, computational and theoretical perspectives*. Cambridge University Press.

Kac, M. 1992. *Grammars and Grammaticality*. John Benjamins Publishing.

Kac, M. 1994. A nonpsychological realist conception of linguistic rules. In Susan D. Lima, Roberta L. Corrigan & Gregory K. Iverson (eds.) *The Reality of Linguistic Rules*, pp. 43–50. John Benjamins Publishing Company.

Kaplan, D. 1990. Words. *Aristotelian Society Supplementary Volume* LXIV: 93–119.

Kaplan, D. 2011. Comments and criticisms: Words on words. *Journal of Philosophy*, 108(9): 504–529.

Kaufman, D. 2002. Composite objects and the abstract/concrete distinction. *Journal of Philosophical Research*, 27: 215–238.

Katz, J. 1965. The relevance of linguistics to philosophy. *Journal of Philosophy*, 62(20): 590–602.

Katz, J. 1971. *Linguistic Philosophy: The underlying reality of language and its philosophical import*. George Allen and Unwin.

Katz, J. 1978. Effability and translation. In Guenthner, F., & Guenthner-Reutter, M. (eds.), *Meaning and Translation: Philosophical and linguistic approaches*, pp. 191–234. Duckworth.

Katz, J. 1981. *Language and Other Abstract Objects*. Rowman.

Katz, J. 1985. An outline of Platonist grammar. In Katz, J., (ed.) *The Philosophy of Linguistics*, pp. 18–48. Oxford University Press.

Katz, J., & Postal, P. 1991. Realism vs. conceptualism in linguistics. *Linguistics & Philosophy*, 14(5): 515–554.

Katz, J. 1996. The unfinished Chomskyan Revolution. *Mind & Language*, 11(3): 270–294.

Katz, J. 1998. *Realistic Rationalism*. Cambridge, MA: The MIT Press.

Kaufman, D. 2002. Composite objects and the abstract/concrete distinction. *Journal of Philosophical Research*, 27: 215–238.

Kim, J. 1998. *Mind in a Physical World*. Cambridge, MA: The MIT Press.

Kincaid, H. 2008. Structural Realism and the Social Sciences. *Philosophy of Science*, 75(5): 720–731.

Kincaid, H. 2013. Introduction: Pursuing a naturalistic metaphysics. In Ross, D., Ladyman, J., & Kincaid, H. *Scientific Metaphysics*, pp. 1–26. Oxford University Press.

Kirby, S. 2013. Language, culture and computation: An adaptive systems approach to biolinguistics. In C. Boeckx, & K. Grohmann (eds.), *Cambridge Handbook of Biolinguistics*. (Cambridge Handbooks in Language and Linguistics). Cambridge University Press.

Kitano, H., & Oda, K. 2006. Self-extending symbiosis: A mechanism for increasing robustness through evolution. *Biological Theory*, 1: 61–66.

Kitcher, P. 1989. Explanatory unification and the causal structure of the world. In: Kitcher P., & Salmon, W. (eds.) *Scientific Explanation*. University of Minnesota Press.

Kocaleva, M., Stojanov, D., Stojanovik, I., & Zdravev, Z. 2016. Pattern recognition and natural language processing: State of the art. *TEM Journal* 5(2): 236–240.

Korbmacher, J., & Schiemer, G. 2018. What are structural properties? *Philosophia Mathematica*, 26(3): 295–323.

Kornai, A., & Pullum, G. 1990. The X-bar theory of phrase structure. *Language* 66(1): 24–50.

Kretzschmar, W. 2015. *Language and Complex Systems*. Cambridge University Press.

Kuipers, T. 2007. *General philosophy of science: Focal issues*. Elsevier.

Labov, W. 2001. *Principles of Linguistic Change, volume 2: Social factors*. Malden: Blackwell Publishers.

Labov, W. 2010. *Principles of Linguistic Change, volume 3: Cognitive and cultural factors*. Wiley-Blackwell.

Labov, W. 2012. *Dialect Diversity in America: The Politics of Language Change*. University of Virginia Press.

Ladusaw, W. 1985. A proposed distinction between Levels and Strata. In The Linguistic Society of Korea (ed.), *Linguistics in the morning calm: Selected papers from the Seoul International Conference on Linguistics*, pp. 37–51. Seoul: Hanshin Publishing Co.

Ladyman, J. 1998. What is structural realism? *Studies in the History and Philosophy of Science* 29(3): 403–424.

Ladyman, J. 2011. Structural realism versus standard scientific realism: the case of phlogiston and dephlogisticated air. *Synthese* 180: 87–101.

Ladyman, J. 2017. An Apology for Naturalized Metaphysics. In Slater, M., & Yudell, Z., (eds.), *Metaphysics and the Philosophy of Science*, pp. 141–161. Oxford University Press.

Ladyman, J., & Ross, D. 2007. *Everything Must Go: Naturalized Metaphysics*. Oxford University Press.

Ladyman, J., & Wiesner, K. 2020. *What is a Complex System?* Yale University Press, New Haven.

Lakoff, G. 1991. Cognitive versus generative linguistics: how commitments influence results. *Language & Communication*, 11 (1/2): 53–62.

Lakoff, G., & Johnson, M. 1999. *Philosophy in the Flesh*. New York: Basic Books.

Lambek. J. 1972. Deductive systems and categories III. In *Lecture Notes in Mathematics*, 274: 57–82.

Landry, E. 2007. Shared Structure Need not be Shared Set-Structure. *Synthese*, 158(1): 1–17.

Langendoen, T. 2003. Merge. In A. Carnie, H. Hayley & M.Willie (eds.), *Formal approaches to function in grammar: in honor of Eloise Jelinek*, pp. 307–318. Amsterdam: John Benjamins.

Langendoen, T., & Postal, P. 1984. *The vastness of natural languages*. Hoboken: Blackwell Publishers.

Lappin, S., Levine, R. & Johnson, D. 2000. The structure of unscientific revolutions. *Natural language & linguistic theory* 18(3): 665–671.

Laudan, L. 1981. A confutation of convergent realism. *Philosophy of Science*, 48: 19–49.

Leitgeb, H. 2020. On non-eliminative structuralism. Unlabeled graphs as a case study, Part A. *Philosophia Mathematica* 28 (3):317–346.

Leitgeb, H., & Ladyman, J. 2008. Criteria of Identity and Structuralist Ontology. *Philosophia Mathematica* 16: 388–396.

Lenneberg, E. 1967. *Biological foundations of language*. New York: John Wiley and Son.

Levelt, W. 1974. *Formal Grammars in Linguistics and Psycholinguistics: Vol. III, Psycholinguistic Applications.* Amsterdam-Philadelphia: John Benjamins.

Lewis, D. 1969. *Conventions: A Philosophical Study*. Harvard University Press.

Lewis, D. 1970. General Semantics. *Synthese* 22 (1/2): 18–67.

Lewis, D. 1975. Languages and language. In Keith Gunderson (ed.), *Minnesota Studies in the Philosophy of Science*, pp. 3–35. University of Minnesota Press.

Liang, P., & Potts, C. 2015. Bringing machine learning and compositional semantics together. *Annual Reviews of Linguistics*, 1(1): 355–376.

Linnebo, O. 2008a. Structuralism and the Notion of Dependence. *The Philosophical Quarterly* 58 (230): 59–79.

Linnebo, O. 2008b. Compositionality and Frege's Context Principle. MS.

Linsky, B., & Zalta, E. 1995. Naturalized Platonism vs. Platonized Naturalism. *Journal of Philosophy* xcii/10: 525–555.

Lobina, D. 2017. *Recursion: a computational investigation into the representation and processing of language*. Oxford: Oxford University Press.

Lopopolo, A., van den Bosch, A., Willems, R.M. 2020. Distinguishing syntactic operations in the brain: Dependency and phrase-structure parsing. *Neurobiology of Language*. Ludlow, P. 2009. Review: Ignorance of Language. *The Philosophical Review*, 118(3): 393–402.

Ludlow, P. 2011. *Philosophy of generative grammar*. Oxford: Oxford University Press.

Ludlow, P. 2014. *Living Words: Meaning Underdetermination and the Dynamic Lexicon*. Oxford: Oxford University Press.

Macarthur, D. 2010. Taking the human sciences seriously. In M. D. Macarthur (ed.), *Naturalism and normativity*. New York: Columbia University Press.

Macarthur, D. 2015. Liberal Naturalism and Second-Personal Space: A Neo-Pragmatist Response to "The Natural Origins of Content". *Philosophia*, 43(3): 565–578.

Maddy, P. 2007. *Second Philosophy: A naturalistic method*. Oxford: Oxford University Press.

Mallory, F. 2020. Linguistic types are capacity-individuated action-types. *Inquiry*, 63 (9–10): 1123–1148.

Malpas, J. 1989. The Intertranslatability of Natural Languages. *Synthese*, 78(3): 233–264.

Marantz, A. 2005. Generative linguistics within the cognitive neuroscience of language. *The Linguistic Review* 22: 429–446.

Marcus, G. 2018. Deep learning: A critical appraisal. Retrieved from arXiv :1801.00631.

Marr, D. 1982. *Vision*. W.H. Freeman and Company: New York.

Martin, M. 1993. Sense modalities and spatial properties. In N. Eilan, R. McCarty, & B. Brewer (eds.) *Spatial representations*, pp. 206–218. Oxford: Oxford University Press.

Martorell, J. 2018. Merging Generative Linguistics and Psycholinguistics. *Frontiers in Psychology*, 9: 1–5.

Matthews, P. 2001. *A Short History of Structural Linguistics*. Cambridge University Press.

Mattson, M. 2014. Superior pattern processing is the essence of the evolved brain. *Frontiers in Neuroscience*, 8(265): 1–17.

Maynes, J., & Gross, S. 2013. Linguistic intuitions. *Philosophy Compass*, 8/8: 714–730.

Meier, T. 2015. *Theory Change and Structural Realism: A General Discussion and an Application to Linguistics*. Dissertation, Ludwig-Maximilians-Universität München.

Menary, R. 2010. Introduction to the special issue on 4E cognition. *Phenomenology ad the Cognitive Sciences*, 9: 459–463.

Millhouse, T. 2019. A simplicity criterion for physical computation. *The British Journal for the Philosophy of Science*, 70(1): 153–178.

Millhouse, T. 2021a. Really real patterns. *Australasian Journal of Philosophy*. DOI: 10.1080/00048402.2021.1941153.

Millhouse, T. 2021b. Compressibility and the reality of patterns. *Philosophy of Science*, 88(1): 22–43.

Miller, G. 2003. The cognitive revolution: A historical perspective. *TRENDS in Cognitive Science*, 7(3): 141–144.

Miller, J. 2021a. A bundle theory of words. *Synthese*, 198: 5731–5748.

Miller, J. 2021b. Words, Species, and Kinds. *Metaphysics*, 4(1): 18–31.

Miller, P. 1999. *Strong Generative Capacity*. CSLI Publications.

Miller, G., & Chomsky, N. 1963. Introduction to the formal analysis of natural languages. In R. Duncan Luce, R. Bush, & E. Galanter (eds.), *The handbook of mathematical psychology* (Vol. II). Wiley.

Millikan, R. 1984. *Language, Thought, and Other Biological Categories*. Cambridge, MA: The MIT Press.

Millikan, R. 2005. *Language: A Biological Model*. Clarendon Press.

Mönnich, U. 2007. Minimalist syntax, multiple regular tree grammars, and direction preserving tree transducers. In J. Rogers & S. Kepser (eds.), *Model Theoretic Syntax at 10. ESSLLI' 07 Workshop Proceedings*, pp. 68–95. Springer.

Montague, R. 1970. Universal grammar. *Theoria*, 36: 373–98. Reprinted in R. Montague, *Formal Philosophy*. Ed. R. Thomason. Yale University Press, 1974, pp. 222–246.

Morgan, M., & Morrison, M. 1999. *Models as Mediators*. Cambridge University Press.

Morrill, G. 2012. Logical Grammar. In Kempson, R., Fernando, T., & Asher, N., (eds.) *The Philosophy of Linguistics*, pp. 63–92. Elsevier.

Moro, A. 2016. *Impossible Languages*. Cambridge, MA: The MIT Press.

Moss, L. 2012. The role of mathematical methods. In Russell, G. & Fara, D. (eds.), *The Routledge Companion to Philosophy of Language*, pp. 533–553. Routledge.

Muller, F. 2010. The characterisation of structure: Definition versus axiomatisation. In Stadler, F. (ed.) *The Present Situation in the Philosophy of Science: The Philosophy of Science in a European Perspective*, vol 1. Springer.

Müller, S. 2018. *Grammatical Theory: From transformational grammar to constraint-based approaches*. Language Sciences Press.

Nefdt, R. 2016. Scientific modelling in generative grammar and the dynamic turn in syntax. *Linguistics & Philosophy*, 39(5): 357–394.

Nefdt, R. 2018a. Languages and other abstract structures. In Christina, C., & Neef, M., (eds.), *Essays on linguistic realism*, pp. 139–184. John Benjamins.

Nefdt, R. 2018b. Structuralism and inferentialism: A tale of two theories. *Logique et analyse*. 61(244): 489–512.

Nefdt, R. 2019a. Infinity and the foundations of linguistics. *Synthese*, 196(5):1671–1711.

Nefdt, R. 2019b. The ontology of words: A structural approach. *Inquiry*, 62(8): 877–911.

Nefdt, R. 2019c. Linguistics as a science of structure. In McElvenny, J. (ed.) *Form and Formalism in Linguistics*, pp. 175–196. Language Sciences Press.

Nefdt, R. 2019d. Why philosophers should do formal semantics (and a bit of syntax too): A reply to Cappelen. *Review of Philosophy & Psychology*, 10(1): 243–256.

Nefdt, R. 2020a. A puzzle concerning compositionality in machines. *Minds & Machines*, 30: 47–75.

Nefdt, R. 2020b. Formal semantics and applied mathematics: An inferential account. *Journal of Logic, Language & Information*, 29(2): 221–225.

Nefdt, R. 2021. Structural realism and generative linguistics. *Synthese*, 199: 3711–3737.

Nefdt, R. (forthcoming). *The Philosophy of Theoretical Linguistics: A contemporary outlook.* Cambridge University Press.

Nefdt, R., & Baggio, G. (2023). Notational Variants and Cognition: The case of dependency grammar. *Erkenntnis.* https://doi.org/10.1007/s10670-022-00657-0

Newen, A., De Bruin, L., & Gallagher, S. 2018. 4E cognition: Historical roots, key concepts, and central issues. In A. Newen, L. De Bruin, & S. Gallagher (eds.). *The Oxford Handbook of 4E Cognition,* pp. 4–16. Oxford University Press.

Newmeyer, F. 1996. *Generative Linguistics: A historical perspective.* Routledge.

Núñez, R., Allen, M., Gao, R., Rigoli, C., Relaford-Doyle, J., & Semenkus, A. 2019. What happened to cognitive science? *Nature Human Behaviour* 3: 782–791.

Odenbaugh, J. 2008. Models. In Sarkar, S., & Plutynski, A. (eds.), *Blackwell Companion to the Philosophy of Biology.* Blackwell Press.

O'Malley, M., & Dupré, J. 2005. Fundamental issues in systems biology. *BioEssays,* 27: 1270–1276.

O'Malley, M., & Dupré, J. 2007. Size doesn't matter: Towards an inclusive philosophy of biology. *Biology & Philosophy,* 22: 155–191.

O'Malley, M., & Soyer, O. 2012. The roles of integration in molecular systems biology. *Studies in History and Philosophy of Biological and Biomedical Sciences,* 43: 58–68.

Paluszek, M., & Thomas, S. 2020. *Practical MATLAB Deep Learning: A project-based approach.* Springer Press.

Parsons, C. 1980. Mathematical intuition. *Proceedings of the Aristotelian Society,* 80: 145–168.

Parsons, C. 1990. The structural view of mathematical objects. *Synthese,* 84(3): 303–346.

Parsons, C. 2004. Structuralism and Metaphysics. *The Philosophical Quarterly,* 54(214): 56–77.

Pelletier, J. 2001. Did Frege believe Frege's Principle? *Journal of Logic, Language, and Information* 10: 87–114.

Peregrin, J. 2008. An inferentialist approach to semantics: time for a new kind of structuralism? *Philosophy Compass,* 3/6: 1208–1223.

Peregrin, J. 2015. *Inferentialism: Why Rules Matter.* Palgrave Macmillan.

Pereplyotchik, D. 2017. *Psychosyntax: The Nature of Grammar and its Place in the Mind.* Springer, Netherlands.

Pi, Y., Liao, W., Liu, M., & Lu, J. 2008. Theory of cognitive pattern recognition. In *Pattern Recognition Techniques, Technology and Applications,* pp. 433–462. Vienna, Austria: InTech.

Piaget, J. 1976. Piaget's Theory. In B. Inhelder et al. (eds.), *Piaget and His School.* Springer-Verlag, New York Inc.

Pietroski, P. 2018. *Conjoining Meanings: Semantics Without Truth Values.* Oxford: Oxford University Press.

Pinker, S. 1997. *How the mind works.* W. W. Norton & Company.

Pitt, D. 2018. What kind of science is linguistics? In Behme, C., & Neef, M. *Essays on Linguistic Realism,* pp. 7–20. John Benjamins, Amsterdam.

Pollard, C., Sag, I. 1994. *Head-Driven Phrase Structure Grammar.* Chicago: University of Chicago Press.

Poibeau, T. 2017. *Machine translation.* Cambridge, MA: The MIT Press.

Poole, G. 2002. *Syntactic theory*. Great Britain: Palgrave.

Postal, P. 1964. Constituent structure: a study of contemporary models of syntactic description. *International Journal of American Linguistics* 30.

Postal, P. 2003. Remarks on the foundations of linguistics. *The Philosophical Forum*. Vol. 34(3): 233–252.

Postal, P. 2009. The Incoherence of Chomsky's 'Biolinguistic' Ontology. *Biolinguistics* 3(1): 104–123.

Priest, G. 2006. What is Philosophy? *Philosophy*, 81(316): 189–207.

Proudfoot, D., & Copeland, J. 2012. Artificial Intelligence. In Margolis, E., Samuels, R., & Stich, S. (eds.), *The Oxford Handbook of Philosophy of Cognitive Science*, pp. 147–182. Oxford University Press.

Psillos, S. 1999. *Scientific realism: How science tracks truth*. Routledge.

Psillos, S. 2006. The Structure, the Whole Structure, and Nothing but the Structure? *Philosophy of Science* 73(5): 560–570.

Pullum, G. 1983. How many possible human languages are there? *Linguistic Inquiry* 14(3): 447–467.

Pullum, G. 2007. Ungrammaticality, rarity, and corpus use. *Corpus Linguistics & Linguistic Theory*, 3: 33–47.

Pullum, G. 2011. The mathematical foundations of Syntactic Structures. *Journal of Logic, Language & Information*, 20(3): 277–296.

Pullum, G. 2013. The central question in comparative syntactic metatheory. *Mind & Language* 28(4): 492–521.

Pullum, G & Scholz, B. 2001. On the distinction between model-theoretic and generative enumerative syntactic frameworks. In de Groote. P., G. Morril & C. Retoré (eds.), *Logical aspects of computational linguistics: 4th international conference*, pp. 17–43. Berlin: Springer.

Pullum, G., & Scholz, B. 2002. Empirical assessment of stimulus poverty arguments. *The Linguistic Review*, 19(1–2): 9–50.

Pullum, G & Scholz, B. 2007. Tracking the origins of transformational generative grammar. *Journal of Linguistics* 43(3): 701–723.

Putnam, H. 1967. The 'Innateness hypothesis' and explanatory models in linguistics. *Synthese*, 17(1): 12–22.

Pylyshyn, Z. 1991. Rules and representations: Chomsky and representational realism. In Kasher, A. (ed.), *The Chomskyan Turn*, pp. 231–251. Oxford: Blackwell.

Quine, W. 1953. Two dogmas of empiricism. In *From a Logical Point of View*, pp. 20–46. Harvard University Press.

Quine, W. 1972. Methodological reflections on current linguistic theory. In Davidson, D., & Harman, G., (eds.), *Semantics of Natural Language*, pp. 442–454. Dordrecht: D. Reidel.

Rambow, O., & Joshi, A. 1997. A formal look at dependency grammars and phrase structure grammars, with special consideration of word-order phenomena. In Wanner, L., (ed.), *Recent Trends in Meaning-Text Theory*, pp. 167–190. John Benjamins.

Reines, M., & Prinz, J. 2009. Reviving whorf: The return of linguistic relativity. *Philosophy Compass*, 4(6): 1022–1032.

Resnik, M. 1982. Mathematics as science of patterns: Ontology and reference. *Nous*, 15: 529–550.

Resnik, M. 1997. *Mathematics as a Science of Patterns*. Clarendon Press.

Rey, G. 2006. The intentional inexistence of language—but not cars. In Stainton, R. (ed.), *Contemporary Debates in Cognitive Science*, pp. 237–255. Blackwell.

Rey, G. 2020. Explanation first! The priority of scientific over "commonsense" metaphysics". In *Language and Reality from a Naturalistic Perspective*, pp. 299–328. Springer Nature.

Richard, M. 2019. *Meanings as Species*. Oxford University Press.

Rietman, E., Karp, R., & Tuszynski, J. 2011. Review and application of group theory to molecular systems biology. *Theoretical Biology and Medical Modelling*, 8(21): 1–29.

Rogers, J. 1998. A descriptive characterization of tree-adjoining languages. *Proceedings of the 17th International Conference on Computational Linguistics (COLING' 98) and the 36th Annual Meeting of the Association for Computational Linguistics (ACL'98)*.

Rogers, J., & Pullum, G. 2008. Expressive power of the syntactic theory implicit in *The Cambridge Grammar of the English Language*. Paper presented at the annual meeting of the Linguistics Association of Great Britain, University of Essex, September 2008. MS.

Ross, D. 2008. Structural realism and economics. *Philosophy of Science*, 75(5): 732–743.

Ross, J. 1967. *Constraints on variables in syntax*. Ph.D. dissertation. MIT. Ross, D., Ladyman, J., & Kincaid, H. *Scientific Metaphysics*. Oxford University Press.

Rumelhart, D., McClelland, J., & Research Group, P. D. P. (eds.). 1986. *Parallel Distributed Processing: Explorations in the microstructure of cognition: Foundations (Vol. 1)*. Cambridge, MA: The MIT Press.

Saatsi, J. 2017. Replacing recipe realism. *Synthese*, 194: 3233–3244.

Sag, I., Wasow, T. & Bender, E. 2003. *Syntactic Theory: A formal introduction (2nd ed.)*. CSLI Publications.

Sampson, G. 2001. *Empirical Linguistics*. Continuum Press.

Sampson, G. 2006. Grammar without grammaticality. *Corpus Linguistics and Linguistic Theory*, 3(1): 1–32.

Samuels, R., Margolis, E., & Stich, S. 2012. Introduction: Philosophy and cognitive science. In Margolis, E., Samuels, R., & Stich, S. (eds.), *The Oxford Handbook of Philosophy of Cognitive Science*, pp. 1–17. Oxford University Press.

Santana, C. 2016. What is language? *Ergo*, 3(19): 501–523.

Sapir, E. 1929. The Status of Linguistic as a Science. *Language* 5: 209.

Saussure, F. de 1916. *Cours de linguistique générale*. In Bally, C. & Sechehaye, A. (eds.) assisted by A. Riedlinger. Payot. 2nd ed. 1922 (subsequent editions essentially unchanged). English translation, *Course in General Linguistics*, by W. Baskin. Philosophical Library, 1959.

Scholz, B., Pullum, G., Pelletier, J. & Nefdt (2022). Philosophy of linguistics. In *the Stanford Encyclopedia of Philosophy*, https://plato.stanford.edu/entries/linguistics.

Seager, W. 2000. Real patterns and surface metaphysics. In Ross, D., Brook, A., & Thompson, D. *Denett's Philosophy: A comprehensive assessment*. Cambridge, MA: The MIT Press.

Sells, P. 2021. The view from declarative syntax. In Allott, N. Lohndal, T., & Rey, G. (ed.), *Blackwell Companion to Chomsky*, pp. 245–266. Wiley-Blackwell.

Shieber, S. 1985. Evidence against the context-freeness of natural language. *Linguistics & Philosophy* 8: 333–343.

Shieber, S., & Schabes, Y. 1991. Synchronous tree-adjoining grammars. *Technical Reports* (CIS).

Schnitzer, M. 1982. Against effability. *Language & Communication*, 2(2): 183–195.

Shapiro, S. 1997. *Philosophy of Mathematics: Structure and ontology*. Oxford University Press.

Shapiro, S. 2005. Categories, structures, and the Frege-Hilbert controversy: The status of metamathematics. *Philosophia Mathematica*, 3(13): 61–77.

Shapiro, S. 2006. Structure and identity. In MacBride, F. (ed.) *Identity and Modality*, pp. 109–145. Oxford University Press.

Shea, N. 2020. *Representation in Cognitive Science*. Oxford University Press.

Sider, T. 2013. *Writing the Book of the World*. Oxford University Press.

Sider, T. 2020. *The Tools of Metaphysics and the Metaphysics of Science*. Oxford University Press.

Simchen, O. 2018. *Semantics, Metasemantics, Aboutness*. Oxford University Press.

Sinha, C. 2010. Cognitive linguistics, psychology, and cognitive science. *The Oxford Handbook of Cognitive Linguistics*, pp. 1–30. Oxford University Press.

Skinner, B.F. 1957. *Verbal Behavior*. Copley Publishing Group.

Skyrms, B. 2010. *Signals: Evolution, Learning and Information*. Oxford University Press.

Smith, N., & Wilson, D. 1985. What Is a Language? In *Syntax and Language Processing*, pp. 325–340. St Martins Press.

Soames, S. 1984. Linguistics and psychology. *Linguistics & Philosophy*, 7: 155–179.

Soames, S. 2010. *The Philosophy of Language*. Princeton University Press.

Sporns, O. 2013. Structure and function of complex brain networks. *Dialogues in Clinical Neuroscience*, 15: 247–262.

Sporns, O. 2014. Contributions and challenges for network models in cognitive neuroscience. *Nature Neuroscience*, 17: 652e660.

Stabler, E. 1997. Derivational minimalism. In Restoré, C. (ed.), *Logical Aspects of Computational Linguistics*, pp. 68–95. Springer.

Stabler, E. 2011. Meta-meta-linguistics. *Theoretical Linguistics*, 37(1/2): 69–78.

Stabler, E. 2019. Three mathematical foundations for syntax. *Annual Review of Linguistics* 5(1): 243–260.

Stabler, E., & Keenan, E. 2003. Structural similarity within and among languages. *Theoretical Computer Science*, 293: 345–363.

Stainton, R. 2014. Philosophy of linguistics. *Oxford Handbooks Online*.

Stainton, R. 2016. A deranged argument against public languages. *Inquiry*, 59(1): 6–32.

Stanley, J. 2008. Philosophy of language in the twentieth century. In Moran, D., (ed.), *The Routledge Companion to Twentieth Century Philosophy*, pp. 382–437. Routledge.

Stanton, K. 2020. Linguistics and philosophy: Break up song. In Nefdt, R., Klippi, C., Karstens, B., (eds.) *The Philosophy and Science of Linguistics*, pp. 409–436. Palgrave Mcmillan.

Stokhof, M., & van Lambalgen, M. 2011. Abstractions and idealisations: The construction of modern linguistics. *Theoretical Linguistics* 37(1/2): 1–26.

Strevens, M. 2020. *The Knowledge Machine: How irrationality created modern science*. Liveright Publishing.

Stokhof, M., & van Lambalgen, M. 2011. Abstractions and idealisations: The construction of modern linguistics. *Theoretical Linguistics* 37(1/2): 1–26.

Suárez, M. 2004. An inferential conception of scientific representation. *Philosophy of Science*, 71: 767–779.

Sullivan, E. 2019. Understanding from machine learning models. *British Journal of the Philosophy of Science* (forthcoming).

Suñe, A., & Martínez, M. 2021. Real patterns and indispensability. *Synthese*, 198 (5): 4315–4330.

Thatcher, J., & Wright, J. 1968. Generalized finite automata theory with an application to a decision problem of second-order logic. *Mathematical Systems Theory* 2(1): 57–81.

Szabó, Z. 1999. Expressions and their representations. *The Philosophical Quarterly*, 49 (195): 145–163.

Szabó, Z. 2000. *Compositionality*. Routledge.

Theakston, A., Lieven, E., Pine, J., & Rowland. C. 2002. Going, going, gone: The acquisition of the verb 'go'. *Journal of Child Language*, 29: 783–811.

Thomason, R. 1974. *Formal Philosophy: Selected Papers by Richard Montague*. New Haven Press.

Tiede, H., & Stout, L. 2010. Recursion, infinity and modeling. In van der Hulst, H. (ed.), *Recursion and Human Language*, pp. 147–158. Mouton de Gruyter.

Tomalin, M. 2006. *Linguistics and the Formal Sciences*. Cambridge University Press.

Tomalin, M. 2010. Migrating propositions and the evolution of Generative Grammar. In Kibbee, D. (ed.), *Chomskyan (r)evolutions*, pp. 315–337. John Benjamins Publishing Company.

Varela, F., Thompson, E., & Rosch, E. 1991. *The Embodied Mind: Cognitive science and human experience*. Cambridge, MA: The MIT Press.

van Fraasen, B. 1980. *The Scientific Image*. Oxford University Press.

van Fraasen, B. 1991. *Quantum Mechanics: An empiricist view*. Oxford University Press.

van Fraasen, B. 2006. Representation: The problem for structuralism. *Philosophy of Science*, 73(5): 536–547.

van Heijenoort, J. 1967. Logic as calculus and logic as language. *Synthese* 17(3): 324–330.

Vasantha Kandasamy, W. & Smarandache, F. 2009. *Groups as Graphs. arXiv e-prints*.

Vijay-Shanker, K., & D. Weir. 1994. The equivalence of four extensions of context-free grammar. *Mathematical Systems Theory*, 27(6): 511–546.

Weisberg, M. 2007. Three kinds of idealization. *The Journal of Philosophy*, 104(12): 639–659.

Weisberg, M. 2013. *Simulation and Similarity: Using Models to Understand the World*. Oxford University Press.

Weisberg, M. 2016. Modeling. In Cappelen, H., Szabó Gendler, T., & Hawthorne, J., *The Oxford Handbook of Philosophical Methodology*, pp. 262–284. Oxford University Press.

Weyl, H. 1952. *Symmetry*. Princeton Science Library.

Wintner, S. 2010. Formal language theory. In Clark, A., Fox, C., Lappin, S. (eds.) *The Handbook of Computational Linguistics and Natural Language Processing*, pp. 11–42. Wiley-Blackwell.

Wittgenstein, L. 1953. *Philosophical Investigations*. Rhees, R. & Anscombe, G.E.M. (trans.), Oxford: Blackwell.

Whorf, B. Science and linguistics. 1956. In Carrol, J. (ed.) *Language, Thought, and Reality: Selected Writing of Benjamin Lee Whorf*, pp. 207–219. Cambridge, MA: The MIT Press.

Worall, J. 1989. The best of both worlds? *Dialectica*, 43(1/2): 99–124.

Woschitz, J. 2021. Scientific realism and linguistics: Two stories of scientific progress. In Nefdt, R., Klippi, C., Karstens, B. *The Philosophy and Science of Language*, pp. 143–177. Palgrave Mcmillan.

Wright, C. 1983. Frege's conception of numbers as objects. *Critical Philosophy*, 1(1): 97.

Yan, K., & Hricko, J. 2017. Brain networks, structural realism, and local approaches to the scientific realism debate. *Studies in History and Philosophy of Science Part C: Studies in History and Philosophy of Biological and Biomedical Sciences*, 64: 1–10.

Youguo, P., Huailin, S., & Tiancai, L. 2007. The frame of cognitive pattern recognition. *Chinese Control Conference*, pp. 694–696, doi: 10.1109/CHICC.2006.4347565.

Yuan, J., Galbraith, D., Dai, S., Griffin, P., & Stewart, N. 2008. Plant systems biology comes of age. *Trends in Plant Science*, 13(4): 165–171.

Zak, D., & Aderem, A. 2009. Systems biology of innate immunity. *Immunological Reviews*, 227(1): 264–282.

Index